About the Authors

Ann Marie Shaw, an elementary school mathematics teacher for 22 years, currently teaches in Paterson NJ. She received her B.A. from St. Leo College, Florida.

Dr. John J. Sico, Jr. is the Assistant Superintendent of Curriculum and Instruction for the Paterson Public Schools and an Adjunct Professor of Mathematics at William Paterson University.

ISBN No. 973-1-56749-603-1

Instructivision, Inc.
16 Chapin Road, P.O. Box 2004
Pine Brook, New Jersey 07058
tel. 973-575-9992
www.instructivision.com

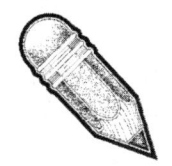

Cluster I

Numbers and Numerical Operations

I-1 Place Value

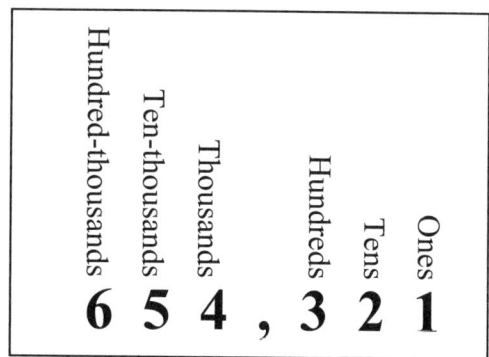

Hundred-thousands	Ten-thousands	Thousands	Hundreds	Tens	Ones
6	5	4 ,	3	2	1

Number facts: When you say a number, it is called a *number*. When you write a number, it is called a *numeral*. The symbols 0, 1, 2, 3, 4, 5, 6, 7, 8, and 9 are called *digits*. Numbers are made up of these ten digits.

5 tens   8 ones

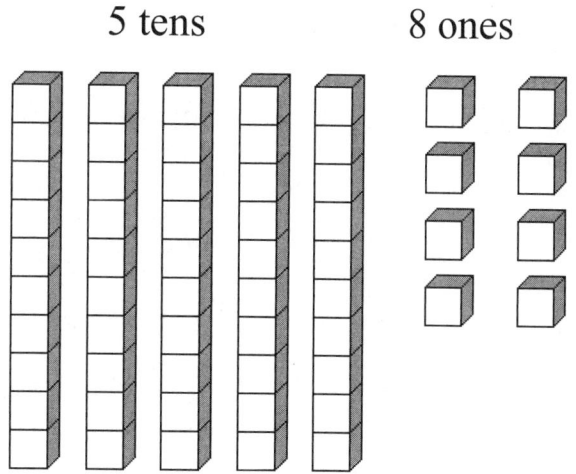

You have 5 groups (or sets) of tens which equal 50 and 8 groups (or sets) of ones which equal 8.

Standard Form: 58
Expanded Form: 50 + 8
Word Form: fifty-eight

Example: What is the place value of 2 in 2̲1?   Answer: tens

What is the place value of 5 in 5̲46?   Answer: hundreds

Now try these:

What is the place value of the 3 in each of the following? Circle the correct answer.

1.   14̲3   hundred-thousands   ten-thousands   thousands   hundreds   tens   ones

2.   3̲10   hundred-thousands   ten-thousands   thousands   hundreds   tens   ones

3.   3̲,184   hundred-thousands   ten-thousands   thousands   hundreds   tens   ones

4.   3̲5,610   hundred-thousands   ten-thousands   thousands   hundreds   tens   ones

I-2 Counting Numbers

$$1, 2, 3, 4, 5, 6, 7, 8, 9, 10, \ldots$$

1.   Can you count to 100? Write all the numbers from 51 to 70.

___   ___   ___   ___   ___   ___   ___   ___   ___   ___

___   ___   ___   ___   ___   ___   ___   ___   ___   ___

Number Facts

The symbols 0, 1, 2, 3, 4, 5, 6, 7, 8, and 9 are called *digits*. Numbers are made up of these ten digits.

2.   Write the largest 2-digit numeral using any of the digits only once.
     Remember that a 2-digit numeral cannot begin with 0.

___   ___

3. Write the <u>smallest</u> 2-digit numeral using any of the digits only once.

   _____ _____

4. Write the <u>largest</u> 3-digit numeral using any of the digits only once. Remember that a 3-digit numeral cannot begin with 0.

   _____ _____ _____

I-3 <u>Whole Numbers</u>

Whole numbers are the counting numbers including the number 0.

$$0, 1, 2, 3, 4, 5, 6, \ldots$$

Now try these:

1. Write the next 10 whole numbers after 6.

   ___ ___ ___ ___ ___ ___ ___ ___ ___ ___

2. What is the next whole number after 101?   _____

3. What is the next whole number after 2,305?   _____

When you count, you are counting by ones. Each counting number is 1 more than the previous number.

$$
\begin{array}{cccccc}
1, & 1+1, & 2+1, & 3+1, & 4+1, & \ldots \\
1 & 2 & 3 & 4 & 5 & \ldots
\end{array}
$$

When you skip count by 2s, each number is 2 more than the previous number. Start at 2 and skip count by 2s.

$$
\begin{array}{cccccc}
2, & 2+2, & 4+2, & 6+2, & 8+2, & \ldots \\
2 & 4 & 6 & 8 & 10 & \ldots
\end{array}
$$

Now try these:

1. Skip count by 2s starting with the number 1. Write the first 10 numerals below.

    <u>1</u> __ __ __ __ __ __ __ __ __

2. Skip count by 2s starting with the number 2. Write the first 10 numerals below.

    <u>2</u> __ __ __ __ __ __ __ __ __

3. Skip count by 2s starting with the number 20. Write the first 10 numerals below.

    <u>20</u> __ __ __ __ __ __ __ __ __

4. Sally looked at the thermometer at the right and told Heather that the temperature was 34°. Heather said, "No, it is not! The temperature is 38°."

    Who is right and why?

    _____

    _____

    _____

    _____

5. Brenda looked at the speedometer below. She told Bill that the arrow is pointing to 41 miles per hour. Bill disagreed with her and said that the arrow is pointing to 42 miles per hour.

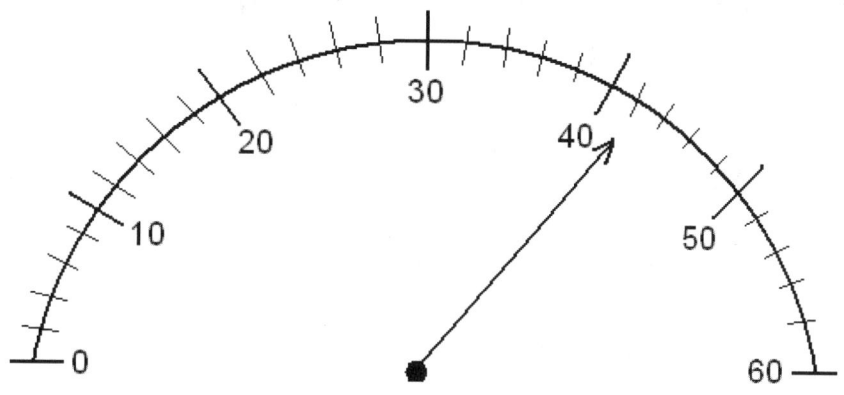

miles per hour

Who is right and why?

_____

_____

_____

6. Which number does point M best represent on the number line?

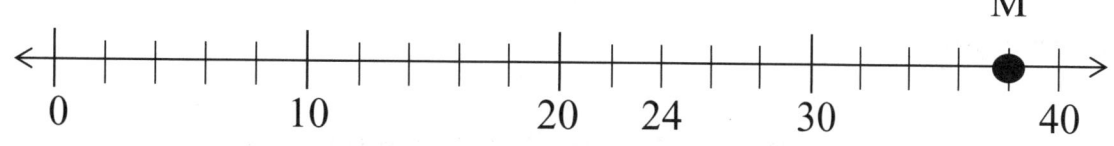

Explain how you arrived at your answer.

_____

_____

7. Look at the number line below and write down the number that represents each point.

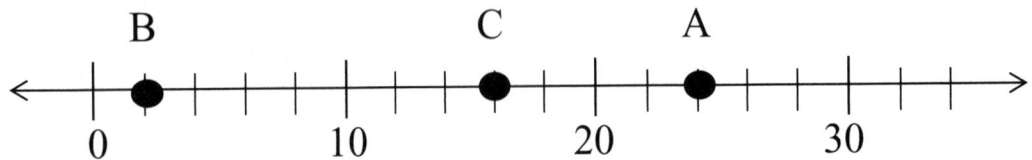

Number _____ represents point A.

Number _____ represents point B.

Number _____ represents point C.

8. Put the following points in the proper place on the number line below.

Point A is represented by the number 18.
Point X is represented by the number 6.
Point Y is represented by the number 12.
Point Z is represented by the number 24.
Point B is represented by the number 48.

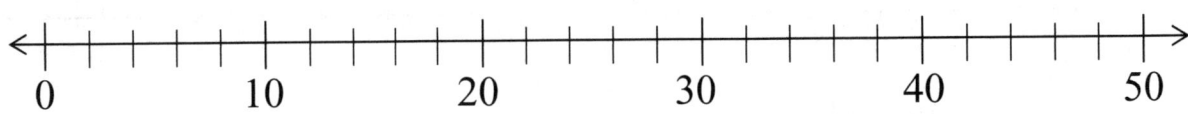

I-4 Math Facts

- Numbers that end in 0, 2, 4, 6, or 8 are *even* numbers.
- Numbers that end in 1, 3, 5, 7, or 9 are *odd* numbers.

Examples: 44 is an even number.
105 is an odd number.
1,079 is and odd number.
12,112 is an even number.

Now try these:

Decide whether or not the number is an even number or an odd number. Circle the correct response for each numeral.

1.	1	even	odd
2.	31	even	odd
3.	461	even	odd
4.	2	even	odd
5.	52	even	odd
6.	314	even	odd
7.	8,615	even	odd
8.	23,372	even	odd

John could only see the ones digit of the numerals below. Decide whether or not the number is an even number or an odd number. Circle the correct response.

9.	XXX,XX4	even	odd
10.	XX1	even	odd
11.	X,XXX,XX7	even	odd
12.	X7	even	odd

Even Numbers	Odd Numbers
0	1
2	3
4	5
6	7
8	9
10	11
12	13
14	15
16	17
18	19
10	21
↓	↓
$x + 2$	$x + 2$
where x is an even number	where x is an odd number

Guess what? You just learned an algebraic expression for even and odd numbers.

1. If you start with an even number and skip count by 2s, will the numbers be even or odd? Show a specific example to support your answer.

_____

_____

2. I am a 2-digit odd number greater than 97.
    What number am I?   _____

3. Write all the odd numbers between 20 and 30.

    _____   _____   _____   _____   _____

4. Write the next five numbers that continue each pattern.

    a. 1, 2, 3, 4,   _____

    b. 2, 4, 6, 8,   _____

    c. 75, 77, 79, 81,   _____

    d. 14, 16, 18, 20,   _____

    e. 1, 3, 5, 7,   _____

    f. 98, 99, 100, 101,   _____

    g. 202, 204, 206, 208,   _____

    h. 55, 57, 59, 61,   _____

    i. 1,000, 1,001, 1,002, _____

I-5 Comparing Numbers

Symbol	Meaning
<	less than
=	equal to
>	greater than

Which is greater: 25 or 24?

Visualize the base 10 blocks as shown:

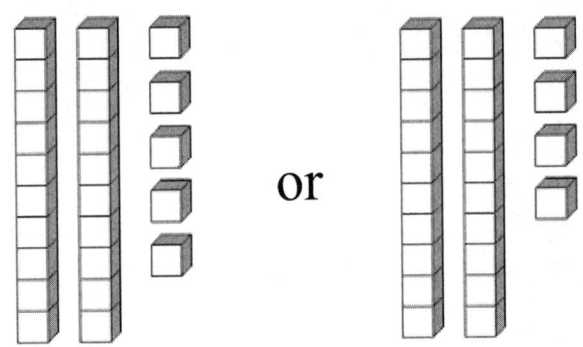

Answer: 25 > 24.

Note: the tens digits are the same, but the ones digits are not the same. Because 5 > 4, 25 > 24. Look at the number line below. The number farther to the right is always the larger number.

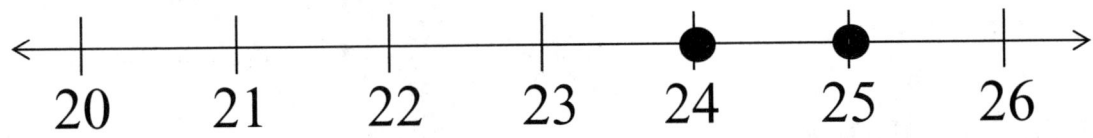

Which is greater: 362 or 345?

Visualize the base 10 blocks:

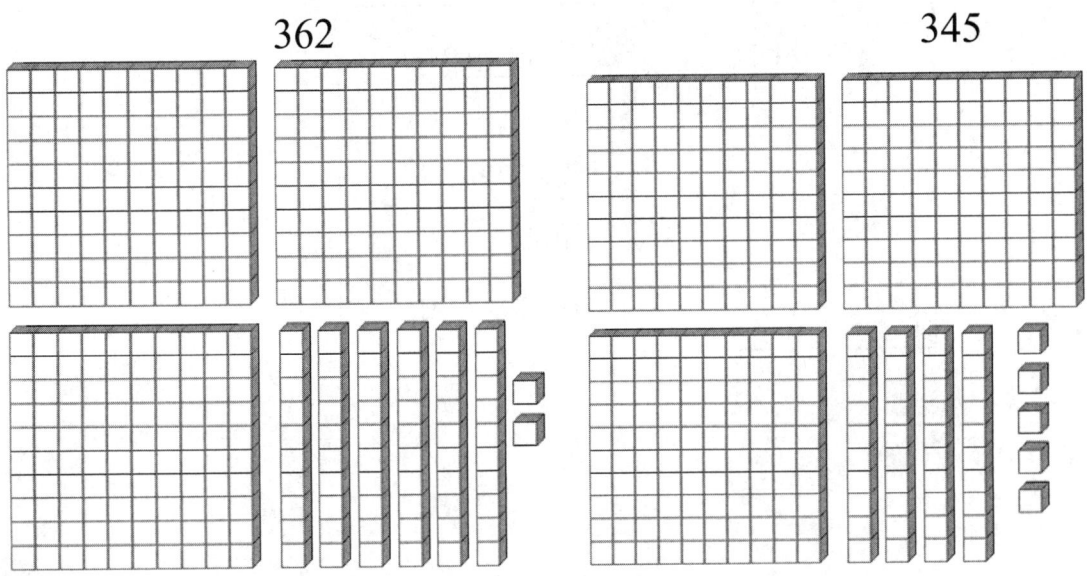

The hundreds digits are the same, but the tens digits are not the same. Because 6 > 4, 362 > 345. Remember to compare the largest place values first.

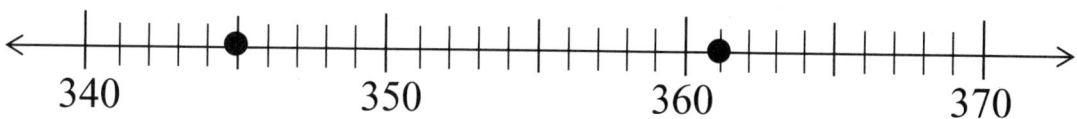

Now try these:

Compare the numbers. Write >, =, or < in each ( ).

1.   235   ( )   229

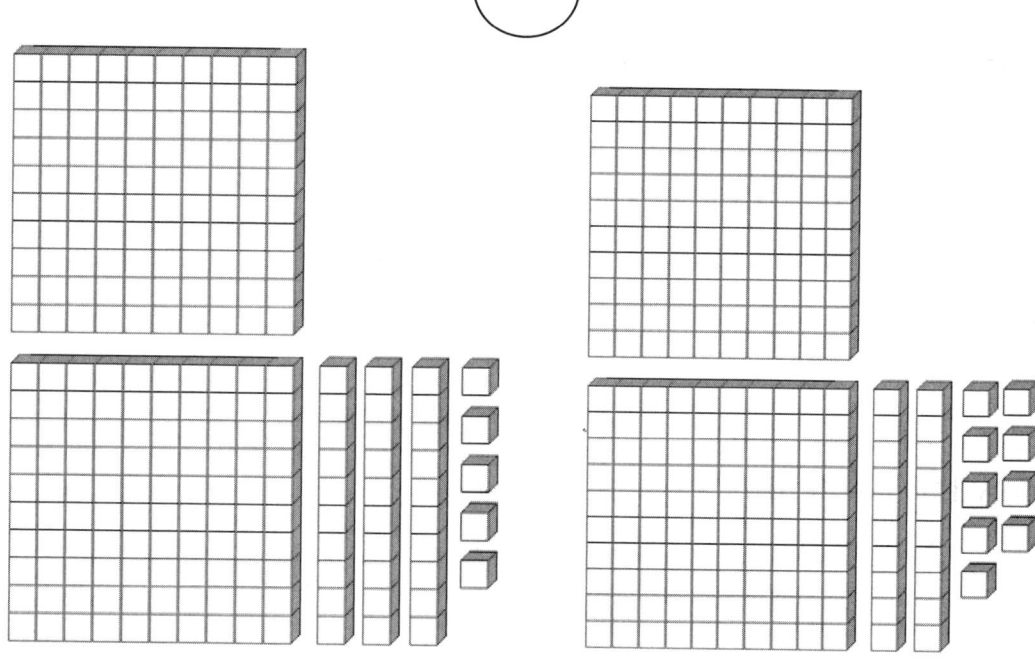

2.   641   ( )   645

3.   1,418   ( )   1,481

4.   439   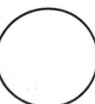   409

5.   3,425      3,542

6.   5,712   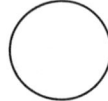   5,699

Math Skills Test

For question 1 to 4, use the picture pattern below.

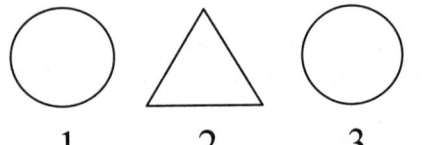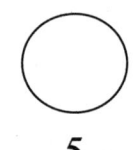

1. Draw the 7th term.

2. Draw the 10th term.

3. Draw the 158th term. (Think: Is 158 an even number or an odd number?)

4. Ahmad could not remember the numbered term that his teacher wanted him to draw. But he knew the number ended in 9. Draw the picture that represents that numbered term.

Fill in the numbers that are missing for each pattern.

5.   288, 290, 292, 294,_____, _____, _____

6.   15, 17, 19, 21, 23,_____, _____, _____, _____

Write the numeral for each number below in standard form and in expanded form.

Word Form	Standard Form	Expanded Form
7. fifty-two	_____	_____
8. two hundred twenty-four	_____	_____
9. seventy-one	_____	_____

For problems 10 to 12 circle the correct answer.

10.   Which word form matches this numeral: 108?

  A. eighteen

  B. one hundred eight

  C. one hundred eighty

  D. one thousand eight

11.   Which word form matches this numeral: 4,305?

  A. four hundred thirty-five

  B. four thousand and three hundred and five

  C. four thousand three hundred five

  D. four thousand thirty-five

12. Which numeral below is two thousand seven hundred fifty-six?

   A. 2,756

   B. 2,000,756

   C. 2,756,000

   D. 27,560

13. Write 105 in word form.

   _____

14. Circle all of the odd numbers.

   82   104   95   3   77   102   8,100

15. Circle all of the even numbers.

   4   7   105   231   18   956   1,002

Use the number below to answer questions 16 to 18. Write the answer on the blank line.

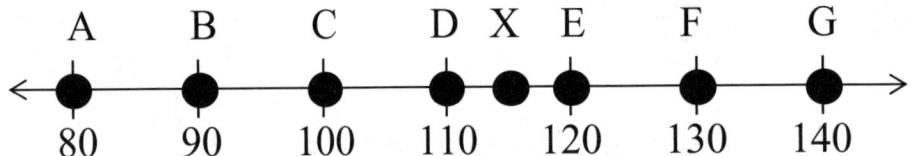

16. Point C is represented by the number _____ on the number line.

17. Point G is represented by the number _____ on the number line.

18. Point X is midway between point D and point _____.

19. Circle the largest number.

    1,842   1,824   1,482   1,284

20. Circle the smallest number.

    4,760   4,067   4,076   4,706

Write the correct symbol in the square: >, =, or <.

21.   184   ☐   148

22.   639   ☐   643

23.   4,984   ☐   4,849

24. Write these numbers in order from smallest to largest.

    156   212   194   704   35

    _____   _____   _____   _____   _____

25. Write these numbers in order from largest to smallest.

    82   67   41   63   58

    _____   _____   _____   _____   _____

26. Match the number in standard form with its word form.

a.   40   _____   (1) eighty five

b.   761   _____   (2) one thousand four hundred seventy six

c.   1,476   _____   (3) forty

d.   85   _____   (4) seven hundred sixty-one

I-6   Writing the Largest and Smallest Odd or Even Numbers

Example 1:   Using the digits 1, 6, and 9, write the largest possible odd number.

Step 1.   _____   _____   _____

Step 2.   1 or 9 must be put in the ones place in order for the number to be odd.

_____   _____   1 or 9

Choose 1 so that you can use the larger digit, 9, in a higher place value.

_____   _____   1

Step 3.   You have the digits 6 and 9 left. Because you want the largest possible odd number, put the 9 in the hundreds place.

9   _____   1

Step 4.   Only the digit 6 is left. Put 6 in the tens place.

9   6   1

The number 961 is the largest possible odd number.

18

Example 2:   Using the digits 1, 5, and 6, write the smallest possible even number

Step 1.   _____   _____   _____

Step 2. 6 is the only even digit. Put 6 in the ones place.

_____   _____   __6__

Step 3. You have the digits 1 and 5 left. Because you want the smallest possible even number, put 1 in the hundreds place.

__1__   _____   __6__

Step 4. Only the digit 5 is left. Put 5 in the tens place.

__1__   __5__   __6__

The number 156 is the smallest possible even number.

Now try these:

1.   Using the digits 1, 2, and 8.

   a.   write the largest possible odd number   _____

   b.   write the smallest possible odd number   _____

   c.   write the largest possible even number   _____

   d.   write the smallest possible even number   _____

2. Using the digits 5, 4, and 3.

    a.   write the largest possible odd number   _____

    b.   write the smallest possible odd number   _____

    c.   write the largest possible even number   _____

    d.   write the smallest possible even number   _____

3. Juan told Alice that if you add the number 2 to any whole number, you will get an odd number. Alice said, "Juan, you are wrong. It depends on what number you started with."

Who is right? Use specific examples to support your answer.

_____

_____

4. Ann's older brother, Ian, is in high school. Ian gave Ann the following clues to figure out his locker combination. Use the clues to find out Ian's locker combination.

Clue Number 1. The 7th number in the pattern is the first number of my locker combination: 18, 20, 22, 24, ____, ____, ____.

Clue Number 2. I am an even number greater than 6 but less than 10. I am the second number in Ian's locker combination.

Clue Number 3. I am the odd number before the 5th number in this pattern: 7, 9, 11, __, __. I am the third number in Ian's locker combination.

Ian's locker combination is: _____, _____, _____.

I-7 Skip Counting by Tens

Can you skip count to 100 by tens?

Write down all the numbers from 10 to 100, skip counting by tens.

__10__ , _____ , _____ , _____ , _____ , __60__ , __70__ , _____ , _____ , _____

Can you skip count to 1,000 by tens?

Write down all the numbers from 10 to 300 skip counting by ten.

__10__ , _____ , _____ , _____ , _____ , _____ , _____ , _____ , _____ , _____

_____ , _____ , _____ , _____ , _____ , _____ , _____ , _____ , _____ , _____

__210__ , _____ , _____ , _____ , _____ , _____ , _____ , _____ , _____ , _____

Rounding to Nearest 10

Let's round numbers to the nearest ten. First, let's answer the following question.

Write down all the numbers between 50 and 60.

50, _____ , _____ , _____ , _____ , _____ , _____ , _____ , _____ , _____ , 60

Now, think of a number line as shown.

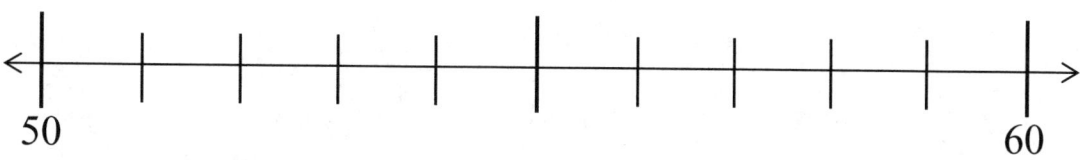

Write down the missing numbers in the boxes below the number line.

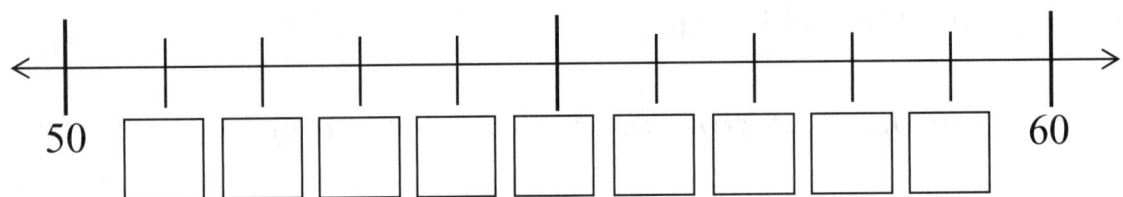

Using the number line you filled in with the missing numbers, answer the following questions. Name the closest ten to:

1. 51 _____

2. 52 _____

3. 53 _____

4. 54 _____

5. 56 _____

6. 57 _____

7. 58 _____

8. 59 _____

Did you notice that the number 55 was missing in questions 1-8? 55 is halfway between 50 and 60? Thus, we need a rule when rounding to the nearest 10.

> Rule: When a number is halfway or more between two tens, round to the greater ten. When the number is less than halfway between two tens, round to the smaller ten.

Since 55 is halfway between 50 and 60, we say 55 rounded to the nearest 10 is 60.

The same rule applies to money.

$$\left.\begin{array}{c} 55¢ \\ 55 \text{ cents} \\ \$0.55 \end{array}\right\} \quad \text{rounds to} \quad \left.\begin{array}{c} 60¢ \\ 60 \text{ cents} \\ \$0.60 \end{array}\right\}$$

Now try these:

Round each number to the nearest ten or nearest ten cents.

1. 28 _____

2. $0.09 _____

3. 34¢ _____

4. $0.43 _____

5. 65¢ _____

6. $0.45 _____

7. 32 _____

8. 102 _____

9. 61 _____

10. $2.58 _____

Rounding to the Nearest 100

Example 1: Round 105 to the nearest hundred.

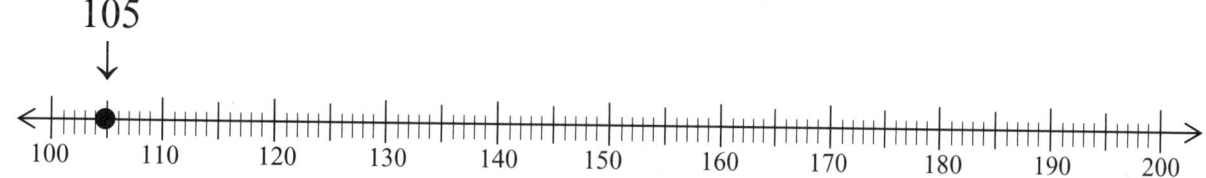

Is 105 closer to 100 or 200? Answer: 105 is closer to 100. Thus, 105 rounded to the nearest hundred is 100.

Example 2: Round 175 to the nearest hundred.

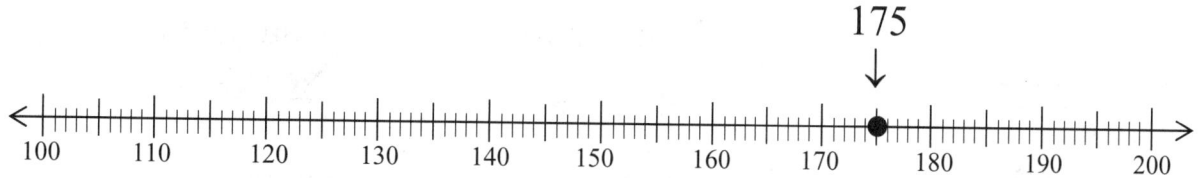

Is 175 closer to 100 or 200? Answer: 175 is closer to 200. Thus, 175 rounded to the nearest hundred is 200.

> **Rule:** When a number is halfway or more between two hundreds, round to the greater 100. When the number is less than halfway between two hundreds, round to the smaller 100.

Examples:

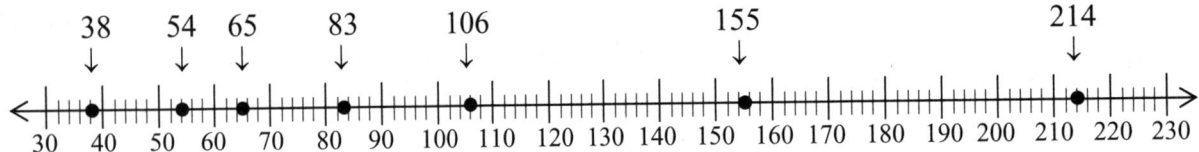

Number	Rounded to Nearest 10	Rounded to Nearest 100
106	110	100
38	40	0
54	50	100
155	160	200
214	210	200
83	80	100
65	70	100

Now try these:

Complete the chart.

	Number	Rounded to Nearest 10	Rounded to Nearest 100
1.	148	_____	_____
2.	72	_____	_____

3.   36   _____   _____

4.   450   _____   _____

5.   81   _____   _____

6.   111   _____   _____

7.   225   _____   _____

8.   222   _____   _____

I-8 Rounding Numbers Continued

<u>Mathematical Rounding Rules for All Whole Numbers</u>

Rule 1. Point to the digit of the place value to which you are rounding off. Circle the digit to the right of this place value. If this circled digit is 0, 1, 2, 3, or 4, replace every digit to the right of the pointed digit with zeroes.

Example 1: Round each of the numbers below to the nearest ten.

a. 14 = 10
   ↑

b. 21 = 20
   ↑

c. 183 = 180
   ↑

d. 133 = 130
   ↑

Example 2: Round each of the numbers below to the nearest hundred.

a. 142 = 100
   ↑

b. 538 = 500
   ↑

c. 729 = 700
   ↑

d. 819 = 800
   ↑

Now try these:

Round each number to the nearest ten.

1. 42 = _____   4. 76 = _____

2. 31 = _____   5. 39 = _____

3. 14 = _____   6. 85 = _____

Round each number to the nearest hundred.

7. 144 = _____

8. 149 = _____

9. 241 = _____

Rule 2.   Point to the digit of the place value to which you are rounding off. Circle the digit to the right of this place value. If this circled digit is 5, 6, 7, 8, or 9, add 1 to the pointed digit and replace every digit to the right of the pointed digit with zeroes.

Example 3: Round each of the numbers below to the nearest ten.

	Hundreds	Tens	Ones	
a. 1⑤ = ↑		1 + 1	0	= 20
b. 2⑦ = ↑		2 + 1	0	= 30

Place Value

Example 4: Round each of the numbers below to the nearest hundred.

Place Value

		Hundreds	Tens	Ones	
a.	1⑤1 =	1 + 1	0	0	= 200
	↑				
b.	4⑥5 =	4 + 1	0	0	= 500
	↑				

More Practice

Round each number to the nearest ten or ten cents.

1. 67 _____

2. 7 _____

3. 75 _____

4. 1 _____

5. 31 _____

6. 311 _____

7. 5 _____

8. 91 _____

9. 15 _____

10. 18 _____

11. 16 _____

12. 92 _____

13. 24¢ _____

14. 93¢ _____

15. 39 _____

16. 94 _____

17. 51 _____

18. 95 _____

19. 11 _____

20. 95 _____

21. 77¢ _____

22. 97 _____

23. 185 _____   24. 98 _____

25. 619 _____   26. 99 _____

27. 20 _____   28. 85¢ _____

I-9 Estimating Sums and Differences

Sometimes you do not need an exact answer, so you *estimate*.

To estimate any two-digit sum, round the addends. Then add.
(Remember: The numbers to be added are called *addends*.)

Round to the nearest 10   36 $\longrightarrow$ 40
Round to the nearest 10 $+\ 23$ $\longrightarrow$ $+\ 20$
                                                           60

Let's compare the estimate to the actual answer. Using a calculator to add
36 and 23, you get:

$$\begin{array}{r} 36 \\ +\ 23 \\ \hline 59 \end{array}$$

This estimate was very close to the actual sum of 59. However, not all
estimates will be this close.

Example:

Round to the nearest 10   34 $\longrightarrow$ 30
Round to the nearest 10 $+\ 44$ $\longrightarrow$ $+\ 40$
                                                           70

Using a calculator to add 34 and 44, you get:

$$\begin{array}{r} 34 \\ +\ 44 \\ \hline 78 \end{array}$$

This estimate of 70 is not very close to the actual sum of 78. Remember that these results are estimates.

Now try these:

Estimate each sum by rounding each addend to the nearest ten and then adding.

1.
$$48 \longrightarrow$$
$$+\,31 \longrightarrow \underline{\hspace{3cm}}$$

3.
$$38 \longrightarrow$$
$$+\,28 \longrightarrow \underline{\hspace{3cm}}$$

2.
$$17 \longrightarrow$$
$$+\,27 \longrightarrow \underline{\hspace{3cm}}$$

4.
$$77 \longrightarrow$$
$$+\,61 \longrightarrow \underline{\hspace{3cm}}$$

Estimate each sum by rounding each addend to the nearest hundred and then adding.

5.
$$147 \longrightarrow$$
$$+\,317 \longrightarrow \underline{\hspace{3cm}}$$

7.
$$860 \longrightarrow$$
$$+\,249 \longrightarrow \underline{\hspace{3cm}}$$

6.
$$450 \longrightarrow$$
$$+\,129 \longrightarrow \underline{\hspace{3cm}}$$

8.
$$481 \longrightarrow$$
$$+\,437 \longrightarrow \underline{\hspace{3cm}}$$

Use the table to answer exercises 9-12.

Liberty Science
Souvenirs Sold

Rubber Snakes	32
Pencils	48
Pens	17

9. To the nearest ten, how many rubber snakes were sold? _____

10. To the nearest ten, how many pencils were sold? _____

11. To the nearest ten, how many pens were sold? _____

12. To the nearest ten, how many souvenirs were sold in all? _____

Estimate each difference by rounding each number to the nearest ten and then subtracting.

Examples:

$$84 \longrightarrow 80$$
$$\underline{-31} \longrightarrow \underline{-30}$$
$$50$$

$$28¢ \longrightarrow 30¢$$
$$\underline{-15¢} \longrightarrow \underline{-20¢}$$
$$10¢$$

Now try these:

1. $38 \longrightarrow$
     $\underline{-24} \longrightarrow$ _____

3. $21 \longrightarrow$
     $\underline{-\ 6} \longrightarrow$ _____

2. $76 \longrightarrow$
     $\underline{-48} \longrightarrow$ _____

4. $84 \longrightarrow$
     $\underline{-22} \longrightarrow$ _____

5.   The temperature in the morning was 69 degrees F. The temperature in the afternoon was 91 degrees F. Which number sentence shows the best way to estimate how many degrees the temperature changed?

   A. $70 + 100 = 170$

   B. $70 + 90\ = 160$

   C. $100 - 70 = 30$

   D. $90 - 70\ = 20$

I-10 Finding Sums

Find the sum.

1. $17 + 0 =$ _____

2. $57 + 8 =$ _____

3. $36 + 8 =$ _____

4. $4 + 5 =$ _____

---

5. $$\begin{array}{r} 90 \\ + 37 \\ \hline \end{array}$$

6. $$\begin{array}{r} 94 \\ + 83 \\ \hline \end{array}$$

7. $$\begin{array}{r} 85 \\ + 94 \\ \hline \end{array}$$

8. $$\begin{array}{r} 23 \\ + 25 \\ \hline \end{array}$$

9. $$\begin{array}{r} 52 \\ + 38 \\ \hline \end{array}$$

10. $$\begin{array}{r} 114 \\ + 26 \\ \hline \end{array}$$

Find the Sum of 3 Numbers:

1. $$\begin{array}{r} 1 \\ 7 \\ + 9 \\ \hline \end{array}$$

2. $$\begin{array}{r} 3 \\ 8 \\ + 2 \\ \hline \end{array}$$

3. $$\begin{array}{r} 10 \\ 12 \\ + 24 \\ \hline \end{array}$$

4. $$\begin{array}{r} 5 \\ 4 \\ + 7 \\ \hline \end{array}$$

5. $$\begin{array}{r} 5 \\ 6 \\ + 2 \\ \hline \end{array}$$

6. $$\begin{array}{r} 20 \\ 40 \\ + 72 \\ \hline \end{array}$$

Adding 2-Digit Numbers

Examples:

$$\begin{array}{r} 25 \\ + 32 \\ \hline 57 \end{array} \qquad \begin{array}{r} {\scriptstyle 1} \\ 47 \\ + 35 \\ \hline 82 \end{array}$$

1. $\begin{array}{r} 23 \\ + 12 \\ \hline \end{array}$

2. $\begin{array}{r} 56 \\ + 62 \\ \hline \end{array}$

3. $\begin{array}{r} 42 \\ + 47 \\ \hline \end{array}$

4. $\begin{array}{r} 26 \\ + 37 \\ \hline \end{array}$

5. $\begin{array}{r} 25 \\ + 14 \\ \hline \end{array}$

6. $\begin{array}{r} 35 \\ + 42 \\ \hline \end{array}$

7. $\begin{array}{r} 29 \\ + 34 \\ \hline \end{array}$

8. $\begin{array}{r} 36 \\ + 47 \\ \hline \end{array}$

9. $\begin{array}{r} 24 \\ + 57 \\ \hline \end{array}$

10. $\begin{array}{r} 42 \\ + 15 \\ \hline \end{array}$

11. $\begin{array}{r} 56 \\ + 52 \\ \hline \end{array}$

12. $\begin{array}{r} 50 \\ + 30 \\ \hline \end{array}$

13. The Eagles scored 25 points in the first quarter. They scored 28 points in the second quarter. How many points did they score in both quarters?

14. The Panthers scored 11 points in the first quarter and 34 points in the third quarter. How many points did they score in both quarters?

15. The Knights scored 34 points in the fourth quarter. The Ghosts scored 46 points. How many points did both teams score in the fourth quarter?

Adding 3 Numbers

1.
```
   14
   40
 + 24
```

2.
```
   32
   30
 + 21
```

3.
```
   21
   63
 + 49
```

4.
```
   33
   17
 + 31
```

5.
```
   28
   25
 + 11
```

6.
```
   52
   61
 + 36
```

7.
```
   11
   12
 + 30
```

8.
```
   32
   10
 + 23
```

9.
```
   16
   18
 + 28
```

10. Sandra bought a pack of gum for 25¢, a pencil for 10¢, and a ruler for 25¢. How much did she spend?

11. Kim went to the movies and bought an ice cream cone for 75¢, popcorn for 60¢, and a soda for 55¢. How much did Kim spend at the movies?

12. Allyson went shopping with her Aunt Christina. She bought a flower for 50¢, a pen for 75¢, and a bow for 32¢. How much did Allyson spend all together?

Adding 3-Digit Numbers

Examples:	$\begin{array}{r} 123 \\ + 456 \\ \hline 579 \end{array}$	$\begin{array}{r} 286 \\ + 108 \\ \hline 394 \end{array}$

Now try these:

1. $\begin{array}{r} 104 \\ + 213 \\ \hline \end{array}$

4. $\begin{array}{r} 720 \\ + 190 \\ \hline \end{array}$

7. $\begin{array}{r} 364 \\ + 458 \\ \hline \end{array}$

2. $\begin{array}{r} 416 \\ + 251 \\ \hline \end{array}$

5. $\begin{array}{r} 214 \\ + 301 \\ \hline \end{array}$

8. $\begin{array}{r} 385 \\ + 178 \\ \hline \end{array}$

3. $\begin{array}{r} 371 \\ + 108 \\ \hline \end{array}$

6. $\begin{array}{r} 441 \\ + 211 \\ \hline \end{array}$

9. $\begin{array}{r} 219 \\ + 134 \\ \hline \end{array}$

I-11 Subtraction

Find the difference.

1. $14 - 6 =$ _____

3. $12 - 4 =$ _____

2. $8 - 7 =$ _____

4. $11 - 2 =$ _____

---

6. $\begin{array}{r} 146 \\ - 32 \\ \hline \end{array}$

8. $\begin{array}{r} 76 \\ - 14 \\ \hline \end{array}$

10. $\begin{array}{r} 56 \\ - 24 \\ \hline \end{array}$

7. $\begin{array}{r} 11 \\ - 3 \\ \hline \end{array}$

9. $\begin{array}{r} 32 \\ - 11 \\ \hline \end{array}$

11. $\begin{array}{r} 76 \\ - 25 \\ \hline \end{array}$

I-12 Solving for the Unknown

Examples: If $? - 14 = 4$, then $? = 4 + 14 = 18$.
  If $? + 6 = 8$, then $? = 8 - 6 = 2$.

1. $8 + 6 = ?$

  $? = $ _____

4. $? - 11 = 11$

  $? = $ _____

2. $10 - 4 = ?$

  $? = $ _____

5. $? - 14 = 3$

  $? = $ _____

3. $9 + 2 = ?$

  $? = $ _____

6. $? - 32 = 4$

  $? = $ _____

I-13 Subtraction of 3-digit Numbers without Borrowing

Examples:	413	718
	− 102	− 605
	311	114

Now try these:

1. 708
  − 104

3. 636
  − 124

5. 759
  − 341

2. 876
  − 123

4. 458
  − 123

6. 914
  − 604

I-14 Subtraction with Borrowing

Example 1.   32 ←—— Minuend
                    - 4 ←—— Subtrahend

Notice the 4 in the ones digit in the subtrahend is greater than the 2 in the minuend.

Visualization:

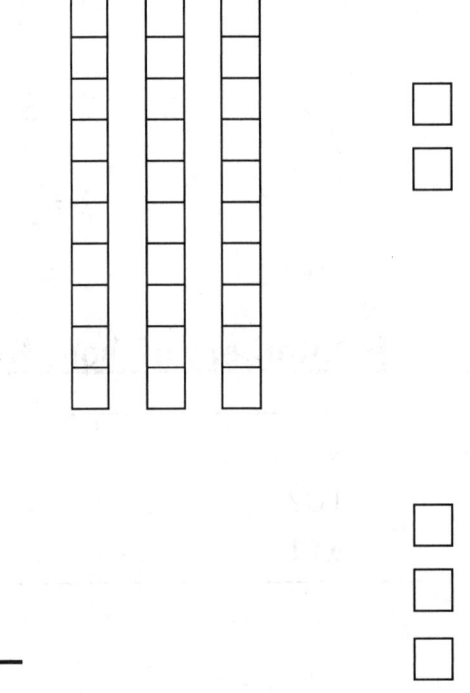

Take one group of ten (called borrowing) and add those ones to the 2 ones you have already.

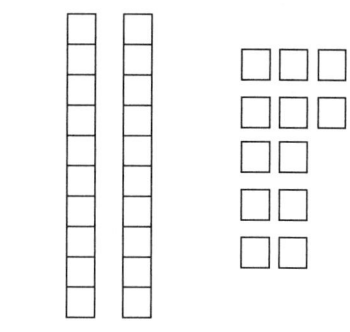

Take 4 away from the group of ones in the minuend. Solution:

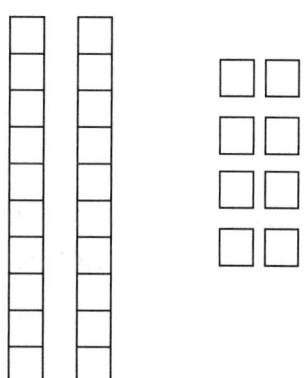

The answer is 28. Let's redo the example without pictures.

$$\begin{array}{r} 2\ _1 \\ \cancel{3}2 \\ -\ 4 \\ \hline 28 \end{array}$$

$$\begin{array}{r} 32 \\ -\ 4 \end{array}$$

Example 2.

$$\begin{array}{r} 25 \\ -\ 7 \end{array}$$

$$\begin{array}{r} 1\ _1 \\ \cancel{2}5 \\ -\ 7 \\ \hline 18 \end{array}$$

Now try these:

1. 27
    − 8

3. 33
    − 14

5. 158
    − 29

2. 82
    − 24

4. 156
    − 29

6. 342
    −128

7. There were 96 adults at the zoo on Monday and 82 adults at the zoo on Tuesday. How many more adults were at the zoo on Monday than on Tuesday?

8. There were 84 children at the movies on Monday and 100 children at the movies on Tuesday. How many more children were at the movies on Tuesday?

9. There were 58 butterfly cocoons in a tree. Thirty-nine butterflies hatched. How many more need to hatch?

10. Jackie counted 56 flowers in the garden. Sandy picked 18. How many flowers were left?

11. Jasmine bought two books. One cost $15.09. The other cost $7.88. How much money did Jasmine spend on books?

12. There were four planes that flew from New York to Florida. The first plane carried 524 passengers. The second carried 142 passengers. The third carried 297 passengers. The fourth plane had 333 passengers. How many passengers flew to Florida?

13. Andrew had 60 days of summer vacation. Each day he did something different for fun. He rode his bike on 18 of the days and played baseball on 30 of the days. He went swimming on all the other days of his vacation. Which number sentence shows one way to find the number of days that Andrew went swimming?

   A. $60 - 18 + 30 =$ ☐

   B. $60 + 18 - 30 =$ ☐

   C. $60 + 18 + 30 =$ ☐

   D. $60 - 18 - 30 =$ ☐

14. What numeral means the same as $40,000 + 1,000 + 50 + 8$?

   A.   41, 508

   B.   41, 058

   C.   40, 518

   D.   40, 158

15. There are 7,904 people in Pine Brook. 4,394 people live in Oakdale. How many more people live in Pine Brook than in Oakdale? Show your work.

16. A baker baked 250 muffins in the morning. By noon, 159 had been sold. How many muffins are left? Show your work.

17. Ally planted 52 seeds, but only 38 plants sprouted. What number sentence shows the number of seeds that didn't sprout?

    A. $52 + 38 = 90$ C. $90 - 52 = 38$

    B. $52 - 38 = 14$ D. $38 - 14 = 24$

18. Carlos scored 50 points in a basketball game, Miguel scored 26, and Toby scored 48. Which number sentence shows how many points Toby and Carlos scored?

    A. $50 + 26 + 48 = 124$ C. $48 + 50 = 98$

    B. $50 - 26 = 24$ D. $48 + 26 = 74$

19. At Madison School there are 54 students in kindergarten, 86 students in first grade, 67 students in second grade, and 58 students in third grade. How many students are in all four grades? Show your work.

20. Jimmy has 56 cars on his racetrack and another 41 cars in a box. How many cars does Jimmy have all together? Show your work.

21. Katelyn's grandmother is 75 years old. Katelyn is 16 years old. How much older is Katelyn's grandmother?

22. Richard bought 5 candy bars. Each candy bar cost $0.65. How much did the 5 bars cost?

23. Emma has 215 cars. Kyle has 879 cars. How many fewer cars does Emma have?

24. Before the start of the soccer game there were 256 people in the stands. Another 244 people arrived during the game. How many people attended the game in all?

25. At the football game on Friday night 785 fans showed up. It was very cold, and 129 fans left early. How many fans stayed until the end of the game?

26. There were 4,149 bees in a beehive in the morning. During the afternoon, 204 bees left the hive. How many bees stayed in the beehive?

I-15 Money

penny	nickel	dime	quarter
1¢	5¢	10¢	25¢

1. Jennifer gave the clerk 2 quarters and 1 dime. How much money did she give the clerk?

2. At the end of the day, Justin had 6 nickels and 4 pennies. How much money did he have?

3. Haley bought popcorn for 90¢. She used 9 coins that were the same. What coins did she use?

   A. pennies   C. dimes

   B. nickels   D. quarters

4. Jason has 5 coins that total 25¢. What are the coins?

   A. pennies   C. dimes

   B. nickels   D. quarters

penny	nickel	dime	quarter
1¢	5¢	10¢	25¢

5. I have 40¢, and all my 3 coins are different. What are the coins?

6. I have 7 coins that total $1.00. What are the coins?

7. Which 3 coins give a value of 30¢?

8. Which 3 coins give a value of 15¢?

9. Which 3 coins give a value of 3¢?

10. Which 3 coins give a value of 16¢?

11. Helena has 6 coins worth $1.50.   How many of each does she have?

____ quarters   ____ dimes   ____ nickels   ____ pennies

12. Bobby has 5 coins equal to 81¢. How many of each does he have?

____ quarters   ____ dimes   ____ nickels   ____ pennies

13. Zane has 4 coins that total 26¢. How many of each does he have?

____ quarters   ____ dimes   ____ nickels   ____ pennies

14. Selena is going to a dance. She bought a new pair of jeans for $15.99 and a top for $9.98. How much did she spend?

15. A baseball glove costs $21.89. You saved $11.40. How much more money will you need to buy the glove?

16. It costs $0.10 to buy a pencil at school. At that price, how much will it cost to buy 15 pencils?

17. Nico has $9.26 saved, and Caroline has $29.62 saved. How much more has Caroline saved than Nico?

18. The soccer team is having a pizza party. They bought 2 cheese pizzas for $6.98, 2 pepperoni pizzas for $9.25, and a vegetable pizza for $11.98. How much money was spent on the pizzas?

19. Wayne went to the Farmer's Market. He bought apples for $2.99, pears for $0.79, and peaches for $1.78. How much money did he spend on the fruit?

20. Kelly bought an apple pie for $7.99. She gave the cashier $20.00. How much change should she get back?

I-16 Exploring Fractions

Definition: A *fraction* is a number that names part of a whole.

Example 1: The large whole square is divided into 4 equal parts. What fractional part of the large square consists of small squares with the letter X in them?

$$\frac{\text{Number of small squares with X}}{\text{Total number of small squares}} = \frac{3}{4} \quad \begin{matrix} \longleftarrow \text{numerator} \\ \longleftarrow \text{denominator} \end{matrix}$$

Example 2: 1 out of 2 equal parts is shaded.

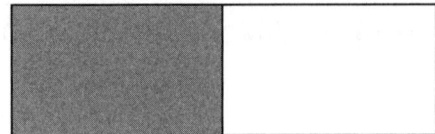

The fraction that tells what part is shaded is $\frac{1}{2}$.

Now try these:

Write the fraction for the part that is shaded. (All parts are equal in questions 1-5.)

1.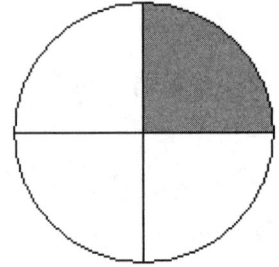

$\dfrac{\text{Numerator}}{\text{Denominator}}$ = _____

2.

$\dfrac{\text{Numerator}}{\text{Denominator}}$ = _____

3.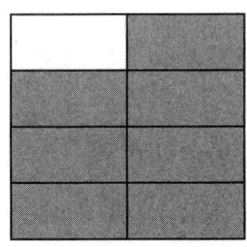

$\dfrac{\text{Numerator}}{\text{Denominator}}$ = _____

4. 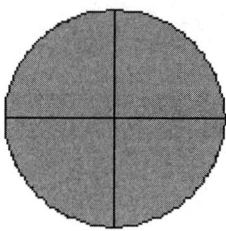 $\dfrac{\text{Numerator}}{\text{Denominator}} =$ _____

5. 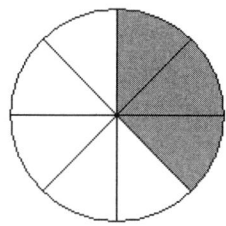 $\dfrac{\text{Numerator}}{\text{Denominator}} =$ _____

6. Circle the correct answer for the shaded area in the figure.

A. $\dfrac{1}{3}$ B. $\dfrac{2}{3}$ C. $\dfrac{3}{3}$ D. cannot be determined

Explain your answer:

_____

_____

_____

7. Write the fraction that tells what part is shaded.

A. _____

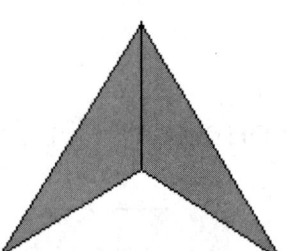

B. _____

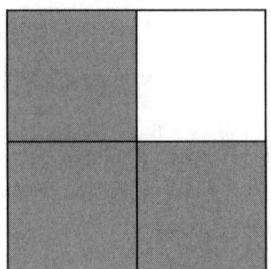

C _____

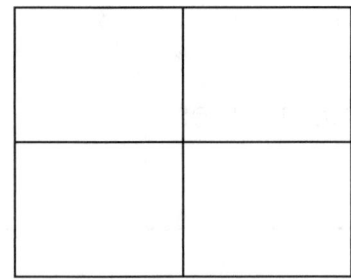

D. _____

8. Shade the part to match the fraction.

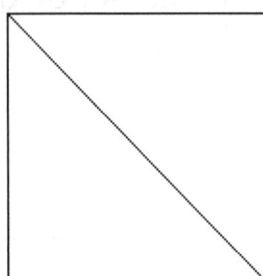

A. $\dfrac{1}{2}$

B. $\dfrac{3}{4}$

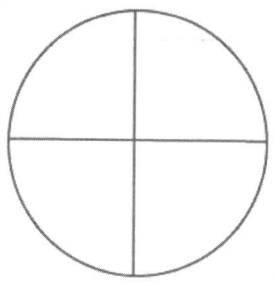

C. $\dfrac{1}{4}$

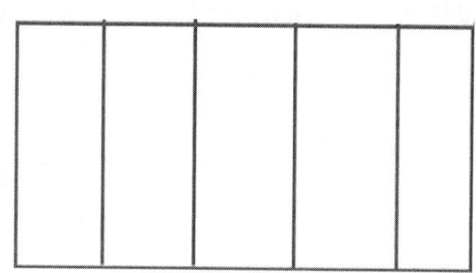

D. $\dfrac{4}{5}$

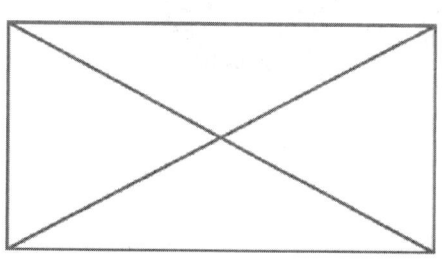

E. $\dfrac{2}{4}$

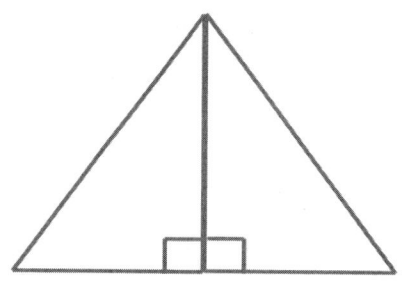

F. $\dfrac{2}{2}$

9. Color the picture to show $\dfrac{2}{5}$ of the bugs are red.

10. What fraction of the coins are quarters? _____

11. Your mother gave you $\frac{3}{5}$ of the 5 dimes in her purse. Draw a picture to show all the dimes you received.

How much money did you get? _____

Equivalent Fractions

Definition: Equivalent fractions have the same amount. Two fractions are equivalent if they have the same amount.

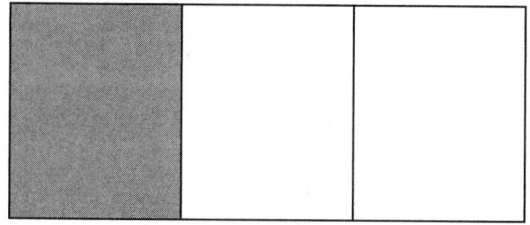 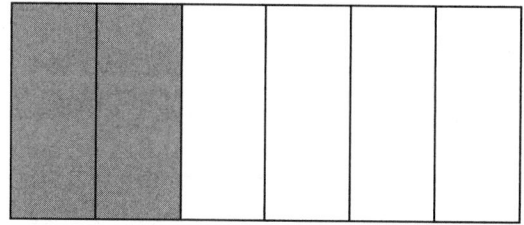

$\frac{1}{3}$ is equivalent to $\frac{2}{6}$.

Now try these:

1. Write the equivalent fraction. Use the figures to help you.

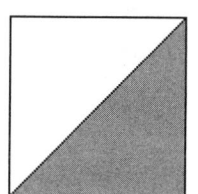 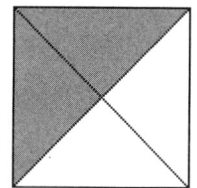   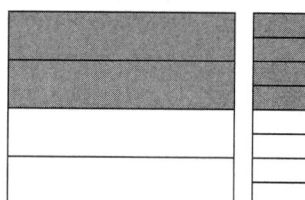 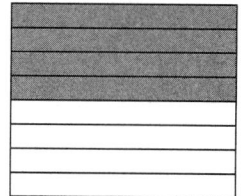

a. $\frac{1}{2}$ _____   b. $\frac{2}{4}$ _____

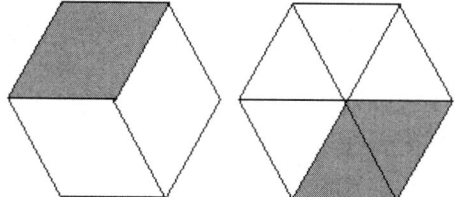   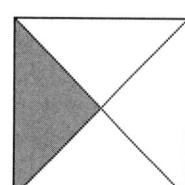 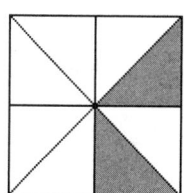

c. $\frac{1}{3}$ _____   d. $\frac{1}{4}$ _____

Use the drawing to help you answer question 2.

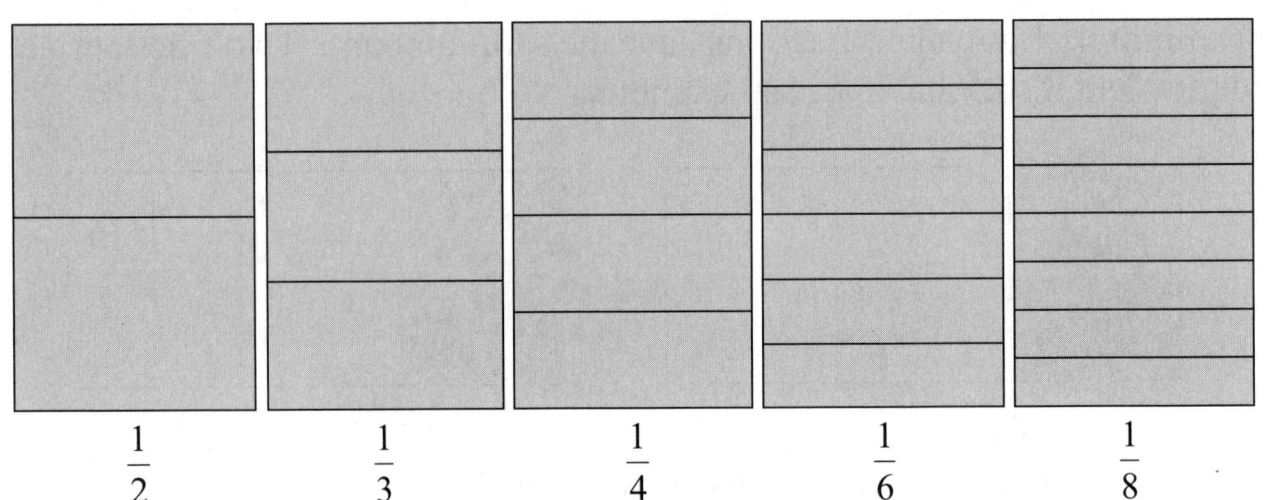

$$\frac{1}{2} \qquad \frac{1}{3} \qquad \frac{1}{4} \qquad \frac{1}{6} \qquad \frac{1}{8}$$

2.   Circle all the equivalent fractions.

a. $\frac{1}{2}$ | $\frac{1}{4}$ $\qquad$ $\frac{2}{4}$ $\qquad$ $\frac{3}{6}$ $\qquad$ $\frac{4}{8}$

b. $\frac{1}{4}$ | $\frac{2}{8}$ $\qquad$ $\frac{2}{3}$ $\qquad$ $\frac{3}{8}$ $\qquad$ $\frac{2}{6}$

c. $\frac{2}{6}$ | $\frac{1}{2}$ $\qquad$ $\frac{1}{4}$ $\qquad$ $\frac{1}{3}$ $\qquad$ $\frac{3}{8}$

d. $\frac{3}{4}$ | $\frac{2}{8}$ $\qquad$ $\frac{1}{2}$ $\qquad$ $\frac{4}{6}$ $\qquad$ $\frac{6}{8}$

3.   Shade $\frac{1}{2}$ of the circle below.

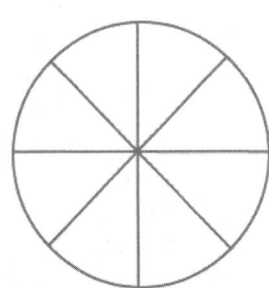

4. Shade $\frac{4}{8}$ of the circle below.

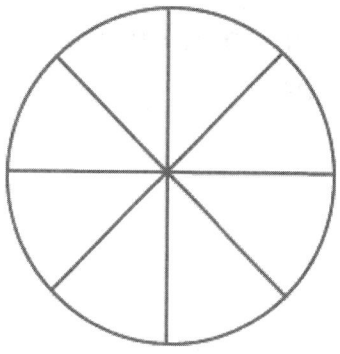

Does $\frac{1}{2} = \frac{4}{8}$? _____

Explain your answer.

_____

_____

_____

_____

5. Thirty students out of 100 students made the honor roll. What fraction tells how many students made the honor roll? Show how you got your answer.

6. There are 10 boys and 9 girls in Ms. Jones' 3rd grade class. What fractional part of the total number of students are boys? Show how you got your answer.

7. What fractional part of the total number of students are girls? Show how you got your answer.

8. Circle all answers that represent less than $\frac{5}{7}$ of the rectangles shaded?

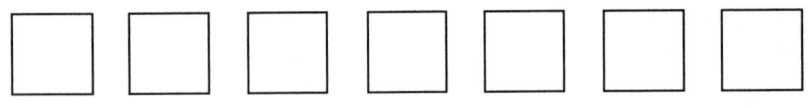

$\frac{1}{7}$ $\frac{2}{7}$ $\frac{3}{7}$ $\frac{4}{7}$ $\frac{5}{7}$ $\frac{6}{7}$ $\frac{7}{7}$

A property of the number 1.

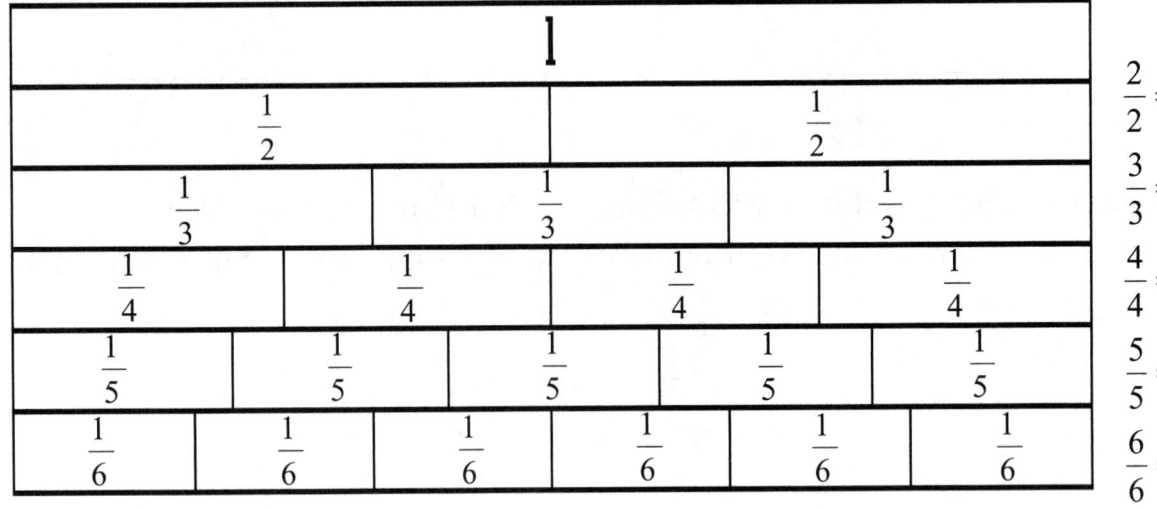

1. How many $\frac{1}{2}$s make 1? _____

2. How many $\frac{1}{3}$s make 1? _____

3. How many $\frac{1}{4}$ s make 1? _____

4. How many $\frac{1}{5}$ s make 1? _____

5. How many $\frac{1}{6}$ s make 1? _____

6. The numerator and denominator of a fraction are equal. Name a whole number that is equivalent to that fraction.

Comparing Fractions

When the denominators in two fractions are the same, compare the numerators to find the greater fraction.

Example: $\frac{2}{4}$ is greater than $\frac{1}{4}$ or $\frac{2}{4} > \frac{1}{4}$

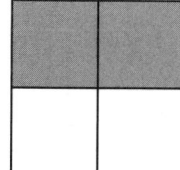 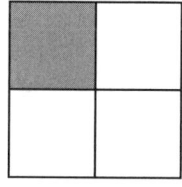

Now try these.

1. Write the fraction that names the shaded part for each sketch. Circle the greater fraction.

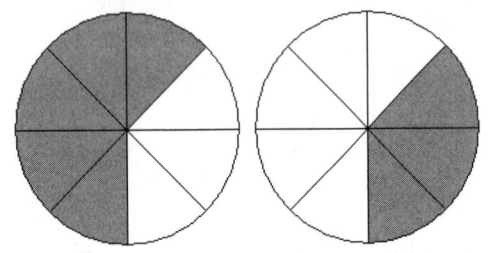 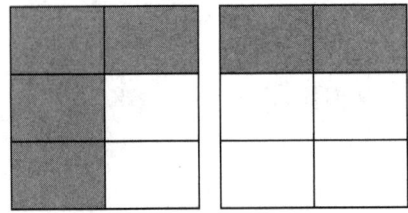

a. _____ _____   b. _____ _____

Use >, <, or = to answer questions 2 - 9. Remember > means "greater than," and > means "less than," and = means "equal to."

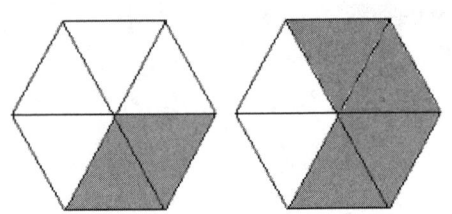

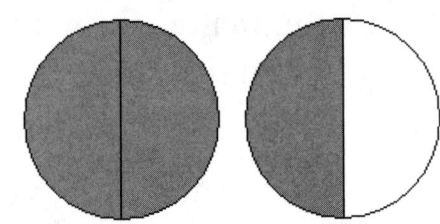

2. $\dfrac{2}{6}$ $\bigcirc$ $\dfrac{4}{6}$ 3. $\dfrac{2}{2}$ $\bigcirc$ $\dfrac{1}{2}$

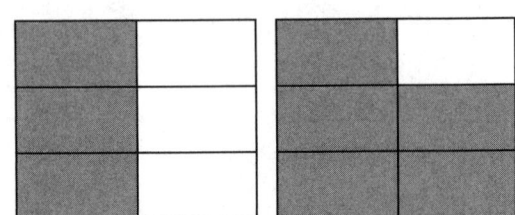

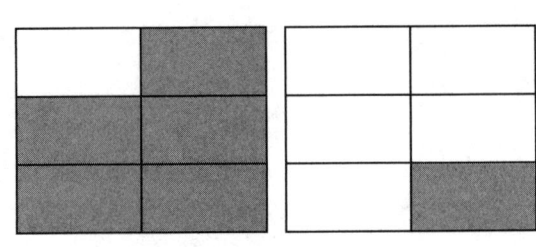

4. $\dfrac{3}{6}$ $\bigcirc$ $\dfrac{5}{6}$ 5. $\dfrac{5}{6}$ $\bigcirc$ $\dfrac{1}{6}$

56

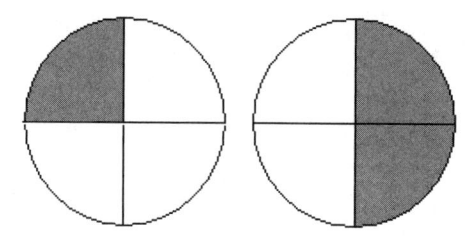

6. $\dfrac{1}{4}$ 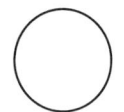 $\dfrac{2}{4}$

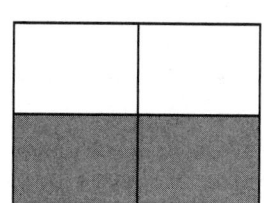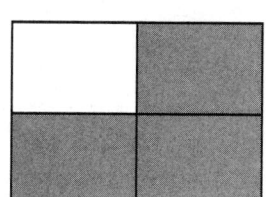

7. $\dfrac{2}{4}$ $\dfrac{3}{4}$

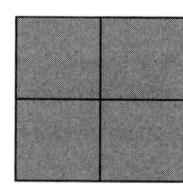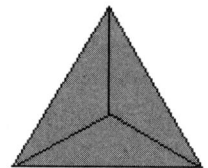

8. $\dfrac{4}{4}$ ◯ $\dfrac{3}{3}$

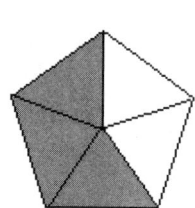

9. $\dfrac{2}{5}$ ◯ $\dfrac{3}{5}$

Use < or > to compare each fraction for questions 10-23.

10. $\dfrac{1}{3}$ 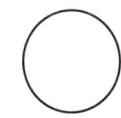 $\dfrac{2}{3}$ 11. $\dfrac{1}{4}$ ◯ $\dfrac{3}{4}$

12. $\dfrac{3}{6}$ ◯ $\dfrac{2}{6}$ 13. $\dfrac{1}{3}$ ◯ $\dfrac{3}{3}$

14. $\dfrac{2}{4}$ ◯ $\dfrac{1}{4}$ 15. $\dfrac{3}{8}$ ◯ $\dfrac{2}{8}$

16. $\dfrac{4}{8}$ ◯ $\dfrac{7}{8}$ 17. $\dfrac{1}{6}$ ◯ $\dfrac{3}{6}$

18. $\frac{6}{8}$ ◯ $\frac{5}{8}$ 19. $\frac{4}{6}$ ◯ $\frac{1}{6}$

20. $\frac{1}{8}$ ◯ $\frac{2}{8}$ 21. $\frac{2}{6}$ ◯ $\frac{5}{6}$

22. $\frac{5}{6}$ ◯ $\frac{1}{6}$ 23. $\frac{2}{2}$ ◯ $\frac{1}{2}$

24. Divide each rectangle below into 10 equal parts.

Shade $\frac{3}{10}$ of the first rectangle and $\frac{5}{10}$ of the second rectangle.

Fill in the square with the correct symbol: <, =, or >.

$$\frac{3}{10} \ \square \ \frac{5}{10}$$

I-17 **Mixed Numbers**

1. There are two equal squares below. (Note: A square has four equal sides and four equal angles.)

   Divide each square into 4 equal parts.

   Shade in all 4 parts of the first square, and shade in 3 parts of the second square.

   a. How many shaded parts are there all together? _____

   b. Can you think of another way of naming the shaded parts? _____

   c. Name the shaded parts another way. _____

2. There are two equal rectangles below. (Note: a rectangle has 4 sides with opposite sides equal and 4 equal angles.)

Divide both rectangles into 5 equal parts.

Shade in all 5 parts of the first rectangle, and shade in 2 parts of the second rectangle.

a. How many shaded parts are there all together? _____

b. Can you think of another way of naming the _____
   shaded parts?

c. Name the shaded parts another way. _____

3. Draw three equal squares.

Divide each square into 4 equal parts.

Shade in all 4 parts of the first square, and shade in all 4 parts of the second square, and shade in 1 part of the third square.

a. How many shaded parts are there all together? _____

b. Can you think of another way of naming the shaded parts? _____

c. Name the shaded parts another way. _____

4. Dave ate $\frac{3}{4}$ of a pizza pie. Juan ate $\frac{1}{4}$ of a pizza. Who ate more pizza?

5. Melissa made an apple pie. She ate $\frac{1}{4}$. Her father ate $\frac{1}{2}$. Who ate the larger share of the pie?

6. David mowed $\frac{1}{4}$ of the lawn before lunch. How much does he have left to mow? Draw a picture to show your answer.

7. Mrs. Chun erased $\frac{1}{2}$ of the blackboard before recess. How much does she have left to erase? Draw a picture to show your answer.

8. Mitchell and Mark shared a pizza. Together they ate $\frac{5}{8}$ of the pizza. How much of the pizza is left? Draw a picture to show your answer.

I-18 **Decimals**

Decimals are numbers that use place value and decimal points.

Example: One tenth can be written as a fraction, $\frac{1}{10}$, or as a decimal, 0.1.

Fraction: $\frac{1}{10}$ Decimal: 0.1   Read: "one tenth"

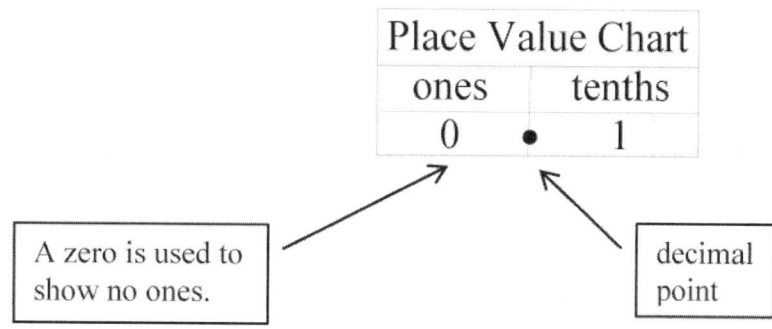

Write a decimal for the part shaded.

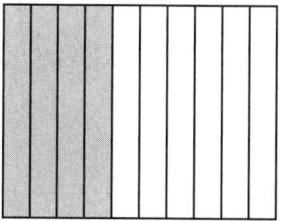

1. _____

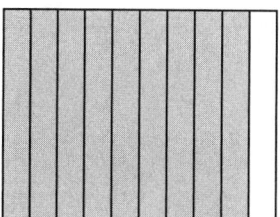

3. _____

2. _____

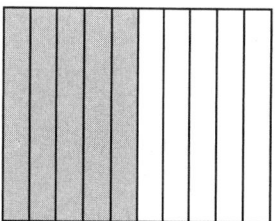

4. _____

Write each fraction as a decimal.

5. $\dfrac{5}{10}$ _____

6. $\dfrac{8}{10}$ _____

7. $\dfrac{3}{10}$ _____

8. $\dfrac{1}{10}$ _____

9. $\dfrac{6}{10}$ _____

10. $\dfrac{4}{10}$ _____

Write each decimal as a fraction.

11. 0.3 _____

12. 0.7 _____

13. 0.1 _____

14. 0.6 _____

15. 0.9 _____

16. 0.4 _____

17. Jason plays basketball. He took 10 shots. He made 5 of these shots. What part of the baskets did Jason make? Write your answer as a fraction and as a decimal.

18. The field hockey team had 10 goals. Kay made 8 of these goals. What part of the goals did Kay make? Write your answer as a fraction and as a decimal.

One hundredth can be written as a fraction, $\frac{1}{100}$, or as a decimal, 0.01.

Example: $\frac{1}{100}$ Read: one hundredth. Written as a decimal: 0.01.

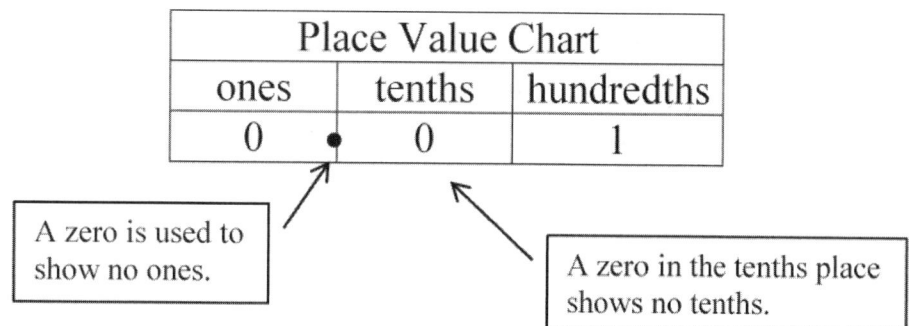

Write each fraction as a decimal.

1. $\frac{7}{100}$ _____

2. $\frac{90}{100}$ _____

3. $\frac{46}{100}$ _____

4. $\frac{18}{100}$ _____

5. $\frac{67}{100}$ _____

6. $\frac{39}{100}$ _____

7. There are 100 students who tried out for the softball team. Only 25 students were selected. What fraction tells how many students were selected?

What fraction tells how many students were not selected?

I-19 Adding and Subtracting Decimals

Example: 1
$$\begin{array}{r} 2.3 \\ +\ 1.8 \\ \hline 4.1 \end{array}$$

Adding: 1. Add the tenths. Regroup if you need. Place a decimal point between the tenths and ones.
          2. Add the ones.

Example: 1   1
$$\begin{array}{r} \cancel{2}.1 \\ -\ 1.5 \\ \hline 0.6 \end{array}$$

Subtracting: 1.   Decide whether to regroup
                2.   Subtract the tenths. Place the decimal point between the tenths and ones.
                3.   Subtract the ones.

Now try these.

Find the sum or the difference.

1.   2.6
          + 1.3

2.   4.4
          - 2.3

3.   3.6
          + 9.2

4.   1.6
          + 1.8

5.   7.1
          - 3.4

6.   4.0
          - 3.6

7. Paula walked 0.6 of a mile on Monday. She walked 1.8 miles on Tuesday. How many miles did she walk all together?

8. Clay ran 1.0 miles on Friday, 1.5 miles on Saturday, and 2.7 miles on Sunday. How many miles did he run in all?

9. Destiny ran 100 meters in 15.2 seconds. Jenna ran 100 meters in 12.9 seconds.

   a. Who ran faster?

   b. How much faster was she?

10. Drew ran the 5K race in 28.36 minutes. It took Cody 53.47 minutes to run the same race.

    a. Who was faster?

    b. How much faster?

11. The temperature is 50.6 degrees outside. Inside, the temperature is 70.8 degrees.

   a. Is it colder inside or outside?

   b. How much colder is it?

I-20 Addition and Multiplication

Example 1:   Add
$$\begin{array}{r} 2 \\ 2 \\ +\,2 \\ \hline 6 \end{array}$$

Multiply

$3 \times 2 = 6$

$2 \times 3 = 6$

Example 2:   Add
$$\begin{array}{r} 5 \\ +\,5 \\ \hline 10 \end{array}$$

Multiply

$2 \times 5 = 10$

$5 \times 2 = 10$

Example 3:   Count the circles:

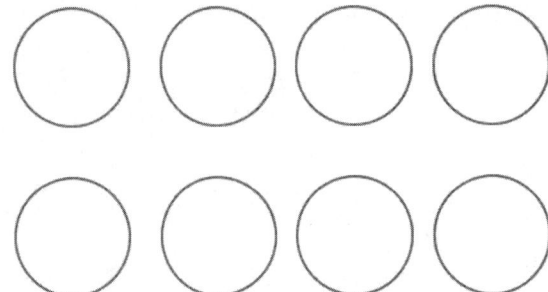

What are rows?

What are columns?

Example:

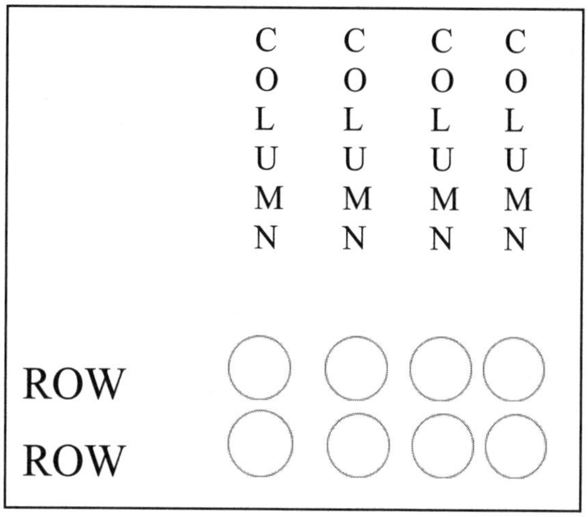

	C O L U M N	C O L U M N	C O L U M N	C O L U M N
ROW	○	○	○	○
ROW	○	○	○	○

How many rows are there? Answer: There are 2 rows.

How many columns are there? Answer: There are 4 columns.

How many circles are there? Answer: $2 \times 4 = 8$. There are 8 circles.

Now try these:

1.　　　4　　　4 fours　　　= _____
　　　　4　　　4×4　　　= _____
　　　　4
　　　 + 4

2.　　　1　　　3 ones　　　= _____
　　　　1　　　3×1　　　= _____
　　　 + 1　　　1×3　　　= _____

3.   6   2 sixes   = _____
$$+\ 6$$
2×6 = _____
6×2 = _____

4.   8   2 eights   = _____
$$+\ 8$$
2×8 = _____
8×2 = _____

5.   7   2 sevens   = _____
$$+\ 7$$
2×7 = _____
7×2 = _____

6.   5   2 fives   = _____
$$+\ 5$$
2×5 = _____
5×2 = _____

7. 

       6   2 sixes   = _____
$$+\ 6$$
2×6 = _____
6×2 = _____

8. Count the small squares.

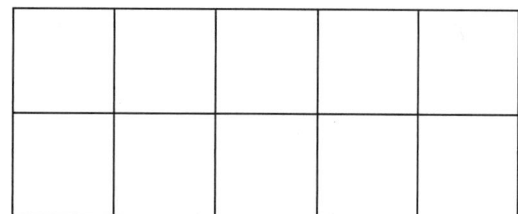

There are _____ small squares.

How many rows are there in the larger rectangle? _____

How many columns are there in the larger rectangle? _____

    5   2 fives   = _____
<u>+ 5</u>   2 × 5   = _____
           5 × 2   = _____

Multiply Using 2 as a Factor

Step 1. Let's review skip counting by two starting with 2.

2, 4, 6, 8, 10, 12, 14, 16, 18, ...

Step 2. Let's make a connection.

Numbering the Terms	Skip Counting by 2	Multiplying by 2
1	2	$2 \times 1 = 2$
2	4	$2 \times 2 = 4$
3	6	$2 \times 3 = 6$
4	8	$2 \times 4 = 8$
5	10	$2 \times 5 = 10$
6	12	$2 \times 6 = 12$
7	14	$2 \times 7 = 14$
8	16	$2 \times 8 = 16$
9	18	$2 \times 9 = 18$

If you can skip count by 2, you can multiply by 2.

Now try these:

1. $\begin{array}{r} 3 \\ \times 2 \\ \hline \end{array}$ 2. $\begin{array}{r} 5 \\ \times 2 \\ \hline \end{array}$ 3. $\begin{array}{r} 2 \\ \times 5 \\ \hline \end{array}$ 4. $\begin{array}{r} 6 \\ \times 2 \\ \hline \end{array}$

5. $\begin{array}{r} 4 \\ \times 2 \\ \hline \end{array}$ 6. $\begin{array}{r} 2 \\ \times 4 \\ \hline \end{array}$ 7. $\begin{array}{r} 1 \\ \times 2 \\ \hline \end{array}$ 8. $\begin{array}{r} 9 \\ \times 2 \\ \hline \end{array}$

9. Anthony put his model cars in 8 rows of 2 each. How many cars does Anthony have?
10. Brittany put her dolls in 6 rows of 2 dolls each. How many dolls does Brittany have?

Multiply Using 3 as a Factor

Step 1. Let's review skip counting by three starting with 3.
3, 6, 9, 12, 15, 18, 21, 24, 27, ...

Step 2. Let's make a connection.

Numbering the Terms	Skip Counting by 3	Multiplying by 3
1	3	$3 \times 1 = 3$
2	6	$3 \times 2 = 6$
3	9	$3 \times 3 = 9$
4	12	$3 \times 4 = 12$
5	15	$3 \times 5 = 15$
6	18	$3 \times 6 = 18$
7	21	$3 \times 7 = 21$
8	24	$3 \times 8 = 24$
9	27	$3 \times 9 = 27$

If you can skip count by 3, you can multiply by 3.

Now try these:

1. 3
   × 2

2. 4
   × 3

3. 5
   × 3

4. 3
   × 8

5. 1
   × 3

6. 3
   × 7

7. 3
   × 9

8. 9
   × 3

9. Ms. Sato has 3 rows and 5 columns of desks in her classroom. How many desks does Ms. Sato have?

10. Roberto has a butterfly collection. He has 3 butterflies in each of 7 rows. How many butterflies does Roberto have?

Multiply Using 4 and 5 as Factors

Let's review skip counting by four starting with 4.

4, 8, 12, 16, 20, 24, 28, 32, 36, ...

Do you see the connection?

$4 \times 1 = 4$	$4 \times 2 = 8$	$4 \times 3 = 12$
$4 \times 4 = 16$	$4 \times 5 = 20$	$4 \times 6 = 24$
$4 \times 7 = 28$	$4 \times 8 = 32$	$4 \times 9 = 36$

If you need to know what 4×5 is and you forget the multiplication fact, skip count by 4 starting with 4 five times: 4, 8, 12, 16, 20. Therefore, $4 \times 5 = 20$.

Lucy asked her teacher, Ms. Sanchez, if she could skip count by 5 starting with 5 four times to find the answer of 4×5. What do you think Ms. Sanchez said?

Let's skip count by 5 four times starting with 5: 5, 10, 15, 20.

$$4 \times 5 = 20$$
$$5 \times 4 = 20$$

Conclusion: $4 \times 5 = 5 \times 4$. You just discovered the commutative property of multiplication.

Let's review:

Complete the table below.

Numbered Term	Skip Count by 2	Skip Count by 3	Skip Count by 4	Skip Count by 5
1	2	3	4	5
2	_____	_____	_____	_____
3	_____	_____	_____	_____
4	_____	_____	_____	_____
5	_____	_____	_____	_____
6	_____	_____	_____	_____
7	_____	_____	_____	_____
8	_____	_____	_____	_____
9	_____	_____	_____	_____

Now try these:

1. $\begin{array}{r} 2 \\ \times\,5 \\ \hline \end{array}$ 2. $\begin{array}{r} 8 \\ \times\,3 \\ \hline \end{array}$ 3. $\begin{array}{r} 9 \\ \times\,4 \\ \hline \end{array}$ 4. $\begin{array}{r} 6 \\ \times\,5 \\ \hline \end{array}$

5. $\begin{array}{r} 9 \\ \times\,5 \\ \hline \end{array}$ 6. $\begin{array}{r} 8 \\ \times\,2 \\ \hline \end{array}$ 7. $\begin{array}{r} 7 \\ \times\,5 \\ \hline \end{array}$ 8. $\begin{array}{r} 8 \\ \times\,4 \\ \hline \end{array}$

Multiply Using 6, 7, 8, and 9 as Factors

Complete the table below.

Numbered Term	Skip Count by 6	Skip Count by 7	Skip Count by 8	Skip Count by 9
1	6	7	8	9
2	_____	_____	_____	_____
3	_____	_____	_____	_____
4	_____	_____	_____	_____
5	_____	_____	_____	_____
6	_____	_____	_____	_____
7	_____	_____	_____	_____
8	_____	_____	_____	_____
9	_____	_____	_____	_____

Now try these:

1. $\begin{array}{r} 8 \\ \times\, 2 \\ \hline \end{array}$
     2. $\begin{array}{r} 8 \\ \times\, 8 \\ \hline \end{array}$
     3. $\begin{array}{r} 9 \\ \times\, 3 \\ \hline \end{array}$
     4. $\begin{array}{r} 7 \\ \times\, 5 \\ \hline \end{array}$

5. $\begin{array}{r} 8 \\ \times\, 9 \\ \hline \end{array}$
     6. $\begin{array}{r} 4 \\ \times\, 7 \\ \hline \end{array}$
     7. $\begin{array}{r} 8 \\ \times\, 8 \\ \hline \end{array}$
     8. $\begin{array}{r} 5 \\ \times\, 6 \\ \hline \end{array}$

Math Fact

☆ 0 has a wonderful multiplication property.

Any number times 0 = 0.

Examples:

1.
$$\begin{array}{r} 4 \\ \times\, 0 \\ \hline 0 \end{array}$$

2.
$$\begin{array}{r} 7 \\ \times\, 0 \\ \hline 0 \end{array}$$

3. $5 \times 8 \times 6 \times 7 \times 0 \times 4 \times 1 = 0$

Math Fact

☆ 1 also has a wonderful multiplication property.

Any number times 1 = that number.

Examples:

1.
$$\begin{array}{r} 6 \\ \times\, 1 \\ \hline 6 \end{array}$$

2.
$$\begin{array}{r} 452 \\ \times\, 1 \\ \hline 452 \end{array}$$

3.
$$\begin{array}{r} 1,856 \\ \times\quad 1 \\ \hline 1,856 \end{array}$$

Now try these:

1. 6×0 = _____

2. 47×0 = _____

3. 185×0 = _____

4. 1×86 = _____

5. 315×1 = _____

6. $8 \times 0 \times 9$ = _____

7. $10 \times 5 \times 8 \times 0$ = _____

8. 152×1 = _____

9. $1 \times 3,756$ = _____

10. $0 \times 56 \times 85 \times 92 \times 541$ = _____

I-21 Comparing Two Fractions When the Denominators are not Equal

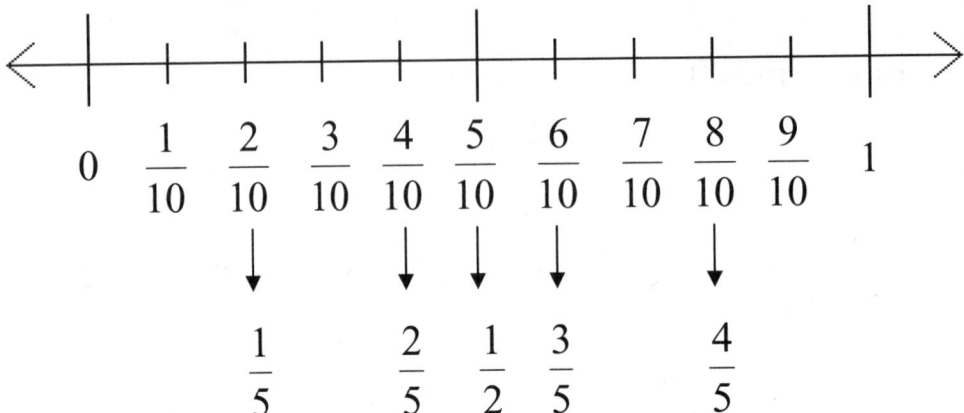

Note: The number farther to the right on the number line is the larger number.

Example 1: Which is greater $\frac{1}{2}$ or $\frac{1}{5}$?

Answer: $\frac{1}{2}$ is farther to the right on the number line than $\frac{1}{5}$.

Thus, $\frac{1}{2} > \frac{1}{5}$.

Let's learn another way to compare fractions with unlike denominators.

Example 2: Which is greater $\frac{1}{2}$ or $\frac{1}{5}$?

Step 1. Starting with the numerator on the left, multiply that numerator by the denominator of the other fraction.

$$\frac{1}{2} \quad \frac{1}{5} \qquad 1 \times 5 = 5$$

Step 2.   Place the 5 under the fraction on the left.

Step 3.   Multiply the numerator on the right by the denominator of the other fraction.

 $1 \times 2 = 2$

Step 4.   Place the 2 under the fraction on the right.

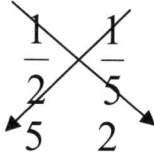

Step 5.   Compare the whole numbers 5 and 2.

$$5 > 2$$

Conclusion: The fraction above the larger whole number is the larger fraction. Therefore, $\dfrac{1}{2} > \dfrac{1}{5}$.

Example 3: Which is greater $\dfrac{1}{3}$ or $\dfrac{1}{4}$?

Solution:

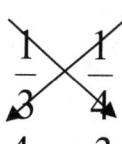

$$4 > 3$$
$$\dfrac{1}{3} > \dfrac{1}{4}$$

Conclusion:

Now try these:

Circle the larger fraction.

1. $\dfrac{1}{5}$ $\dfrac{1}{3}$ 2. $\dfrac{1}{4}$ $\dfrac{1}{2}$

3. $\dfrac{1}{3}$ $\dfrac{1}{6}$ 4. $\dfrac{1}{8}$ $\dfrac{1}{6}$

5. $\dfrac{1}{2}$ $\dfrac{1}{10}$ 6. $\dfrac{1}{5}$ $\dfrac{1}{8}$

7. $\dfrac{1}{6}$ $\dfrac{1}{10}$ 8. $\dfrac{1}{8}$ $\dfrac{1}{10}$

9. $\dfrac{2}{3}$ $\dfrac{1}{4}$ 10. $\dfrac{2}{5}$ $\dfrac{3}{7}$

11. $\dfrac{5}{8}$ $\dfrac{2}{3}$ 12. $\dfrac{6}{8}$ $\dfrac{2}{5}$

13. $\dfrac{20}{30}$ $\dfrac{3}{4}$ 14. $\dfrac{4}{9}$ $\dfrac{3}{8}$

15. $\dfrac{50}{60}$ $\dfrac{7}{8}$ 16. $\dfrac{9}{10}$ $\dfrac{3}{4}$

I-22 Connecting Multiplication and Division

Let's review:

$$\begin{array}{r} 4 \\ \times 2 \\ \hline 8 \end{array} \qquad\qquad 2 \times 4 = 8 \qquad\qquad 4 \times 2 = 8$$

Now let's rewrite the above multiplication problems this way.

$$\begin{array}{r} 4 \\ \times ? \\ \hline 8 \end{array} \qquad\qquad ? \times 4 = 8 \qquad\qquad 4 \times ? = 8$$

What number does the symbol " ? " represent?   Answer: $? = 2$

Guess what? You just did a division ($\div$) problem.

Example 1:   If $5 \times 4 = 20$, then $20 \div 5 = 4$ and $20 \div 4 = 5$.

Example 2:   If $3 \times 6 = 18$, then $18 \div 3 = 6$ and $18 \div 6 = 3$.

Remember: Division is the opposite of multiplication.

Example 3: $35 \div 5 = ?$

Think:	$? \times 5 = 35$
Skip count by 5:	5, 10, 15, 20, 25, 30, 35
Answer:	$5 \times 7 = 35$
Therefore,	$35 \div 5 = 7$

Example 4: $35 \div 7 = ?$

Think:	$? \times 7 = 35$
Skip count by 7:	7, 14, 21, 28, 35
Answer:	$5 \times 7 = 35$
Therefore,	$35 \div 7 = 5$

Now try these:

1. $4 \div 2$ = _____   10. $10 \div 2$ = _____

2. $8 \div 2$ = _____   11. $9 \div 1$ = _____

3. $9 \div 3$ = _____   12. $35 \div 7$ = _____

4. $16 \div 4$ = _____   13. $8 \div 4$ = _____

5. $25 \div 5$ = _____   14. $49 \div 7$ = _____

6. $16 \div 2$ = _____   15. $9 \div 9$ = _____

7. $20 \div 4$ = _____   16. $0 \div 2$ = _____

8. $28 \div 7$ = _____   17. $6 \div 6$ = _____

9. $27 \div 9$ = _____   18. $40 \div 5$ = _____

Multiply a 2-Digit Number × a 1-Digit Number

Example:
$$\begin{array}{r} 42 \\ \times\ 2 \\ \hline 4 \end{array}$$
Multiply 2 by the digit in the ones place. (Skip count by 2 if necessary.) $2 \times 2 = 4$
Place the 4 in the ones column.

$$\begin{array}{r} 42 \\ \times\ 2 \\ \hline 84 \end{array}$$
Multiply 2 by the digit in the tens place. $2 \times 4 = 8$
Place the 8 in the tens column.
Answer: $42 \times 2 = 84$

Now try these:

1.   80
    ×4
	2. 61
	    ×3
		3. 50
		    ×1
			4. 86
			    ×0

5.   62
    ×3
	6. 33
	    ×2
		7. 81
		    ×3
			8. 80
			    ×2

Multiply a 2-Digit Number With Regrouping

Example:

$$35 \times 2$$

Multiply 2 by the digit in the ones place.
$2 \times 5 = 10$

$$\begin{array}{r} 1 \\ 35 \\ \times\ 2 \\ \hline 0 \end{array}$$

Place the 0 in the ones column and put the 1, which represents 10, on top of the 3 in the tens place.

$$\begin{array}{r} 1 \\ 35 \\ \times\ 2 \\ \hline 70 \end{array}$$

Multiply 2 by the digit in the tens place, but remember to add the 1. $2 \times 3 + 1 = 6 + 1 = 7$
Place the 7 in the tens column.
Answer: $35 \times 2 = 70$

Now try these:

1.   25
    ×4
	2. 36
	    ×2
		3. 62
		    ×3
			4. 84
			    ×3

5.   67
    ×5
	6. 84
	    ×7
		7. 36
		    ×4
			8. 63
			    ×6

9.   73
    ×8
	10. 85
	    ×2
		11. 93
		    ×6
			12. 87
			    ×4

13.	97 × 2	14.	85 × 3	15.	95 × 2	16.	17 × 3
17.	64 × 5	18.	49 × 3	19.	64 × 5	20.	72 × 9

Open-ended Questions

21. Rosa has 8 coins in her purse. She does not have any pennies. What is the greatest amount of money she can have?

    What is the least amount of money she can have?

22. You are going to the toy store. You have $10.00. How many different toys can you buy for your $10.00? (You may only buy each toy once.)

Toy Store Prices			
Ball	$2.50	Puzzle	$4.50
Bat	$3.50	Transformer	$3.00
Train	$5.00	Barbie	$7.00

23. Using the numbers 3, 6, 5, and 4:

    What is the largest number you can write?   _____

    What is the smallest number you can write?   _____

24. You have a red, a blue, and a green crayon. How many ways can you arrange the 3 crayons? List your answer.

25. Darlene has 7 coins that total $0.78. What are the coins she has? Draw and explain your answer.

26. Sharon has 4 pieces of gum. Jason has 2 times as many pieces of gum as Sharon. Brian has 3 times as many pieces as Jason. How many pieces of gum does each person have?

27. The Ortiz family is having a pool party. They bought a 5-foot sandwich from the deli. The sandwich was divided into 20 pieces. If there are 4 Ortiz family members and 6 friends of the family at the party, what fraction of the sub can each person have?
Show and explain your answer.

28. Kevin drinks 3 glasses of milk each weekday and 2 glasses on each weekend day. Make a chart to show how much milk Kevin drinks during one week.

29. At the Farmer's Market, 4 grapefruit cost $1.00. How many grapefruits can Diego buy for $4.00?

30. What number can be added to or subtracted from a number and the sum or difference is the number you started with?
Explain and give an example for your answer.

31. Sara drinks 2 to 4 glasses of milk every day. Her coach wanted her to drink 30 glasses of milk. Which is a reasonable number of days it will take Sara to drink 30 glasses of milk?

    A. Fewer than 8 days

    B. Between 8 and 10 days

    C. Between 7 and 16 days

    D. More than 29 days

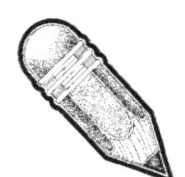

Cluster II

Geometry

II-1 <u>Points and Line Segments</u>

Point A   • A

Line segment $\overline{AB}$

Line $\overleftrightarrow{AB}$

Ray $\overrightarrow{AB}$

Use letters and symbols to name each picture.

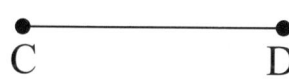   • P

1. _____   2. _____

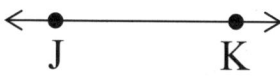

3. _____   4. _____

II-2 Parallel Lines and Intersecting Lines

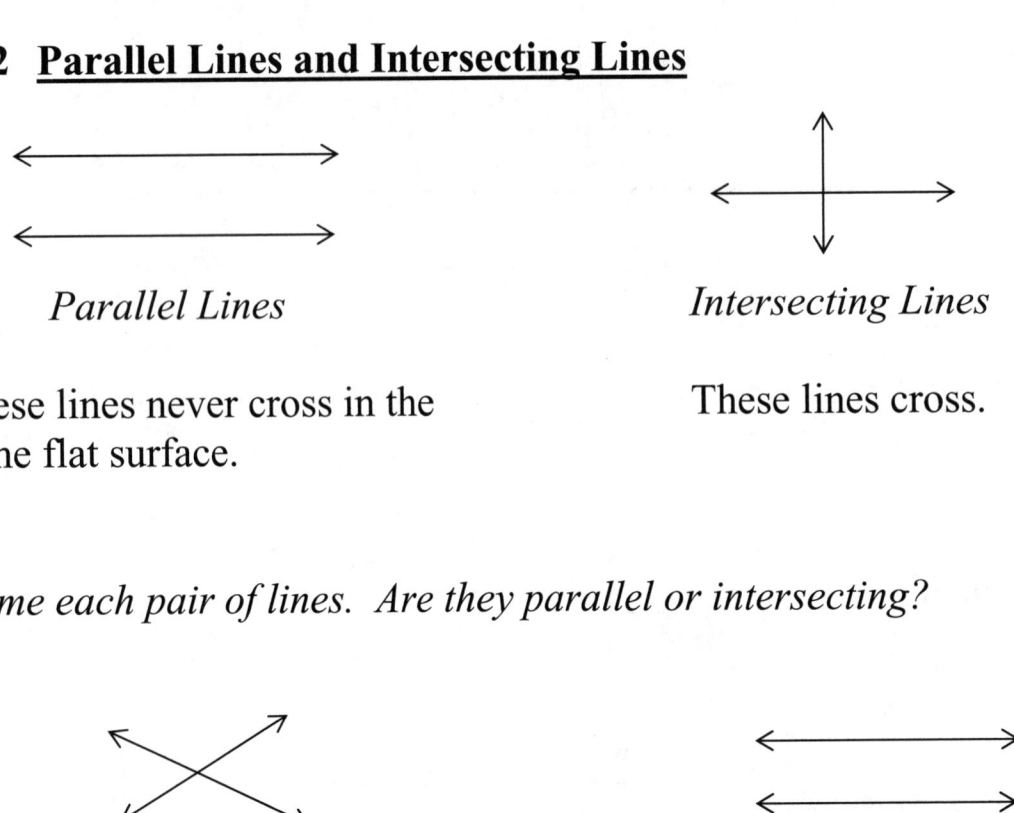

Parallel Lines

These lines never cross in the same flat surface.

Intersecting Lines

These lines cross.

Name each pair of lines. Are they parallel or intersecting?

1. _____

2. _____

3. _____

4. _____

5. _____

6. _____

II-3 Rays and Angles and Perpendicular Lines

A ray has one end point.

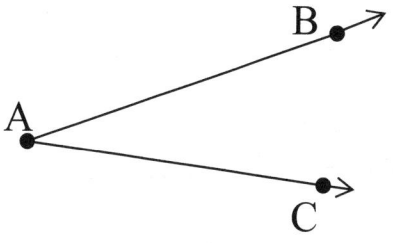

read $\overrightarrow{AB}$

Two rays with the same endpoint make an angle.

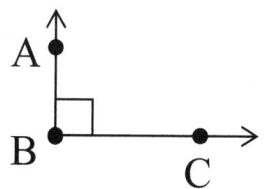

read ∠BAC, ∠A, or ∠CAB

An angle that measures 90° is a right angle.

read ∠ABC, ∠B, or ∠CBA

Use the letters to name each figure.

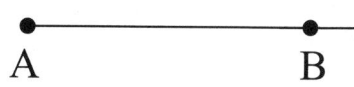

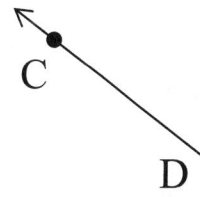

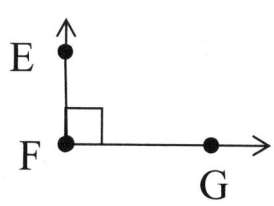

1. _____   2. _____   3. _____

Name each angle 3 different ways.

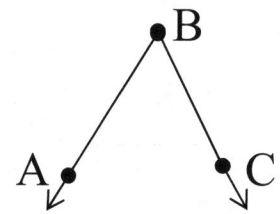

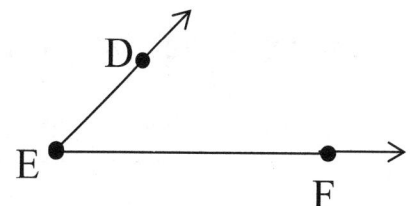

4. _____ _____ _____   5. _____ _____ _____

89

6. Which angles are right angles?

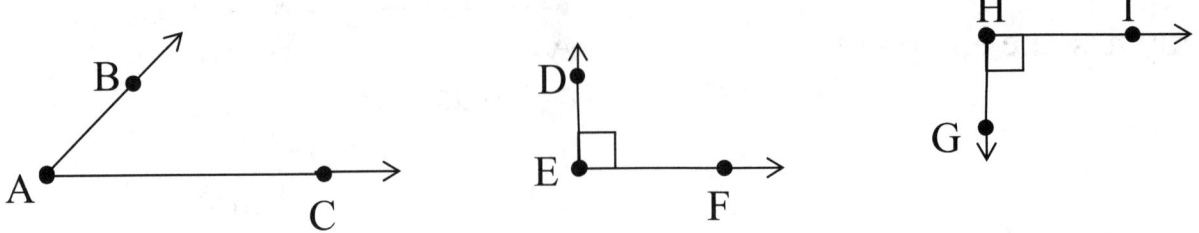

The right angles are: _____

Use the figure to answer questions 7-9.

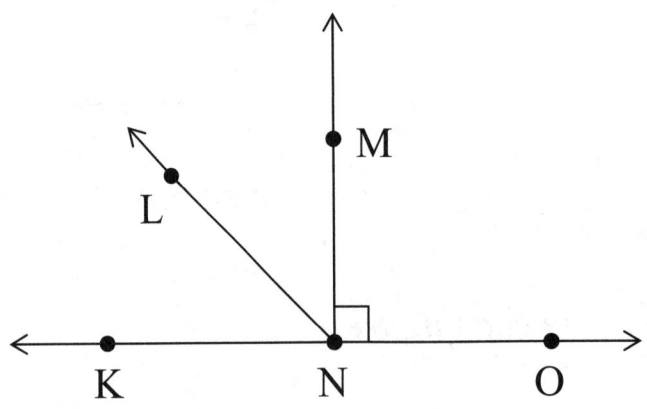

7. Name four rays.

_____ _____ _____ _____

8. Name five angles.

_____ _____ _____ _____ _____

9. Name two right angles.

_____ _____

II-4 **Ordered Pairs**

Ordered pairs tell you how far to count over and then up or down to identify an intersection on the grid.

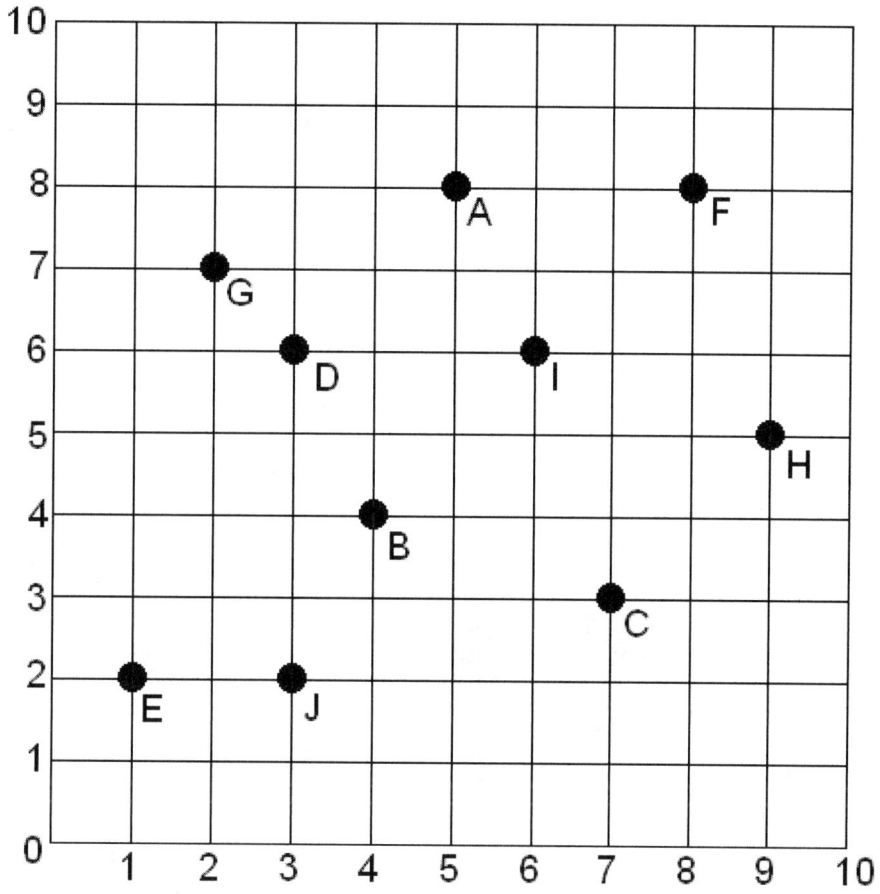

Write the letter at the intersection of each ordered pair.

1. (3, 6) _____

2. (6, 6) _____

3. (7, 3) _____

4. (9, 5) _____

5. (4, 4) _____

6. (1, 2) _____

7. (5, 8) _____

8. (3, 2) _____

9. (8, 8) _____

10. (2, 7) _____

11. Which ordered pair can you use to find the square?

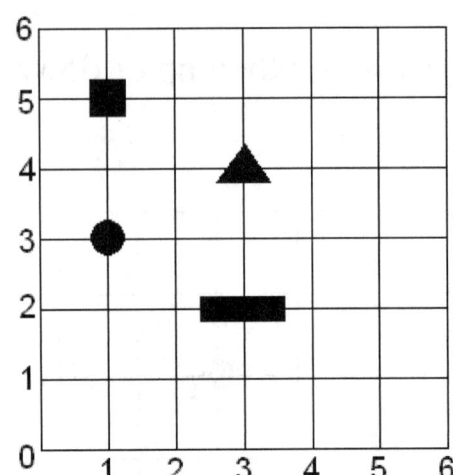

A. (3, 4)

B. (1, 5)

C. (2, 2)

D. (1, 3)

12. What shape is located at the ordered pair (4, 3)?

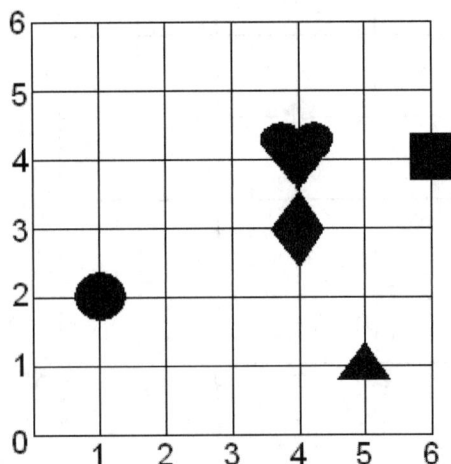

A. heart

B. square

C. diamond

D. triangle

II-5 Congruent Figures

Congruent figures are figures that have the same size and shape.

Example: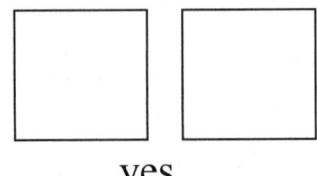
yes

Example: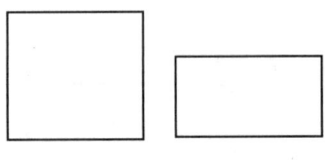
no

Are these figures congruent?

 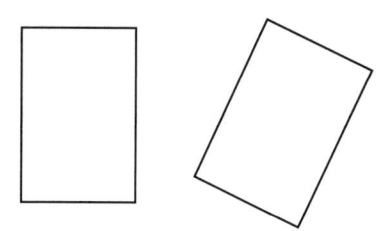

1. _____   2. _____

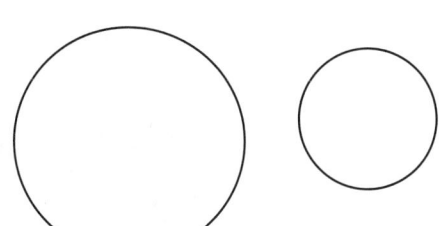

3. _____

4. Circle the two figures in each row that are congruent.

a.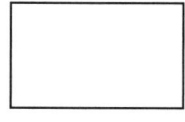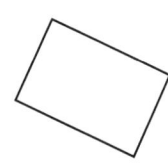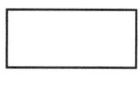

b.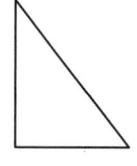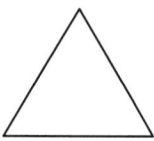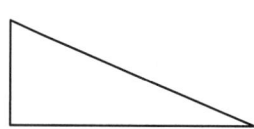

c.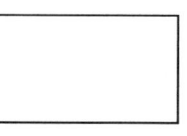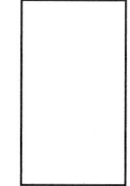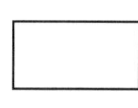

II-6 **Symmetry**

If you can fold a figure in half so that both halves are identical, then the figure is *symmetrical*. The fold is called the *line of symmetry*.

Are these figures symmetrical? (The dash represents the fold.)

_____   _____   _____

_____   _____

Transformation

A transformation is a change in the shape by sliding, flipping, or turning.

Slide: move to a new spot.

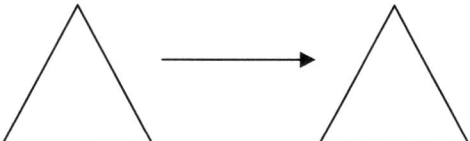

Flip: reverse the shape to its mirror image.

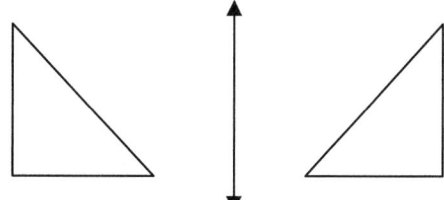

Turn: rotate the shape to face a new direction.

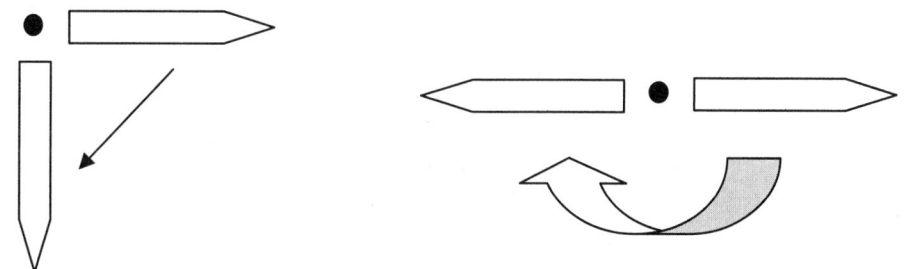

a quarter turn around the point   a half turn around the point

Tell how each figure was moved. Write *slide, flip,* or *turn.*

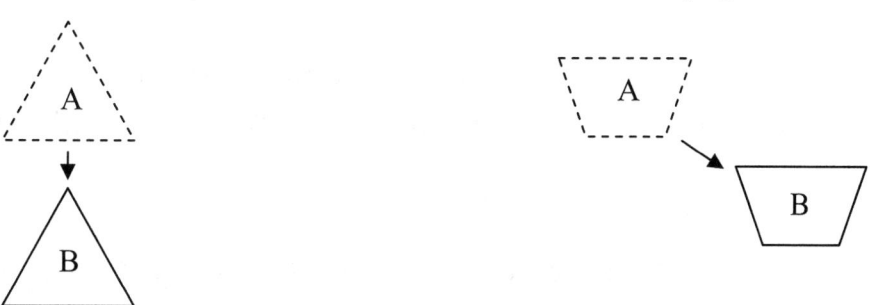

_____   _____

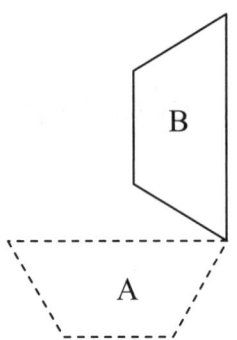

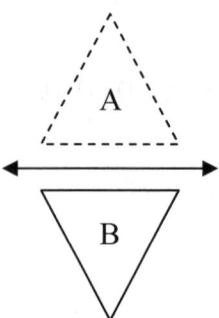

_____   _____

II-7 **Properties of a Circle**

Diameter, Radius, and Center

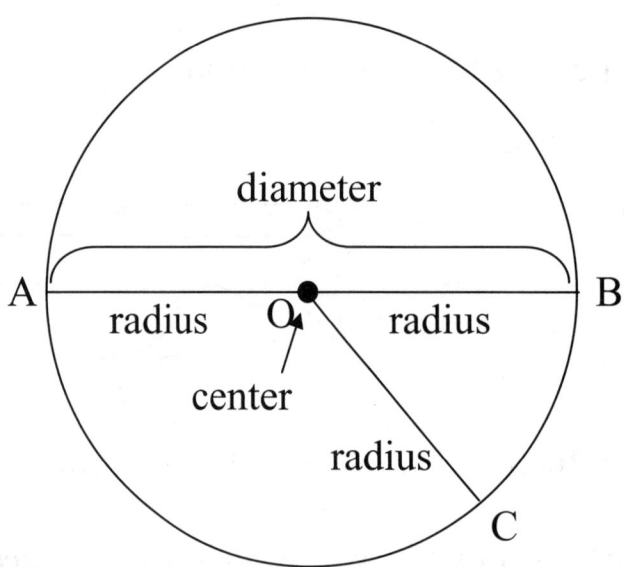

The *diameter* is a line segment from one side of a circle to the other through the center of the circle. $\overline{AB}$ is a diameter of the circle with center O.

A radius is a line segment from the center of a circle to a point on a circle. $\overline{OC}$, $\overline{OB}$, and $\overline{OA}$ are radii of the circle with center O.

Now try these:

1. Write the *diameter* or *radius* for $\overline{AB}$.

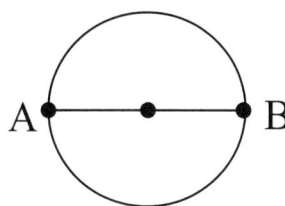 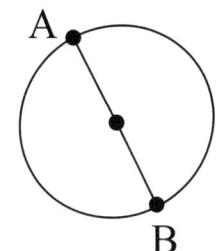 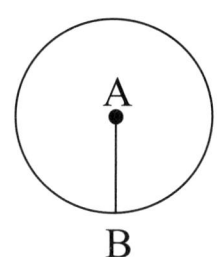

a. _____ b. _____ c. _____

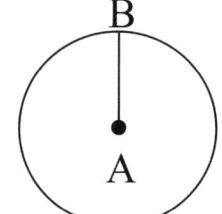 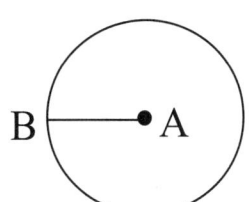

d. _____ e. _____

2. Use the letters to name the diameter and 2 radii in each circle.

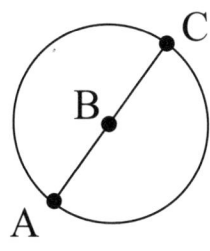 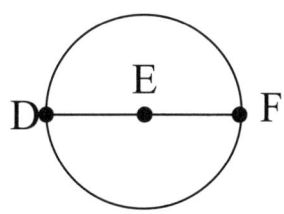 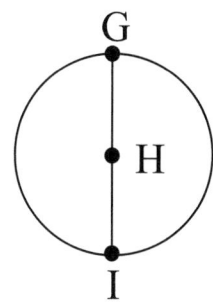

diameter _____ diameter _____ diameter _____

radius _____ radius _____ radius _____

radius _____ radius _____ radius _____

3. Refer to circle 0.

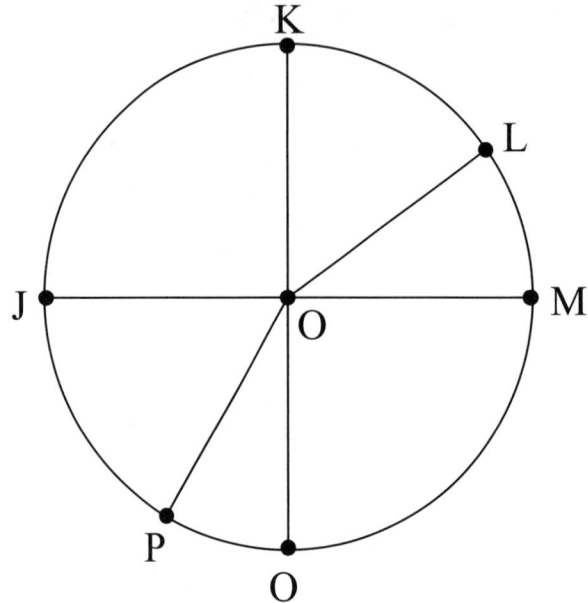

Name the center of circle O above.   _____

Name all radii of circle O above.

_____, _____, _____, _____, _____, and _____.

List the diameters that are shown in circle O above.

_____ and _____

Measurement

Definition: *Perimeter* is the distance around a figure.

2 in.

1 in.   1 in.

2 in.

Add the length of each side to find the perimeter.

2 + 1 + 2 + 1 = 6 in.

Now try these:

Find the perimeter of each figure.

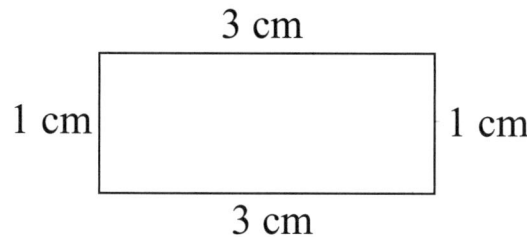

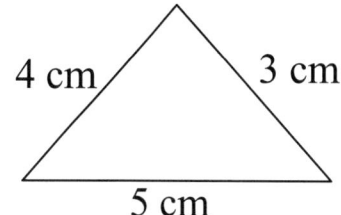

1. _____   2. _____

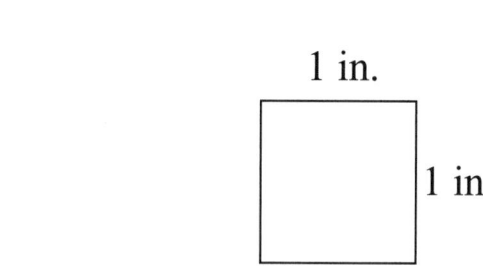

rectangle   square

3. _____   4. _____

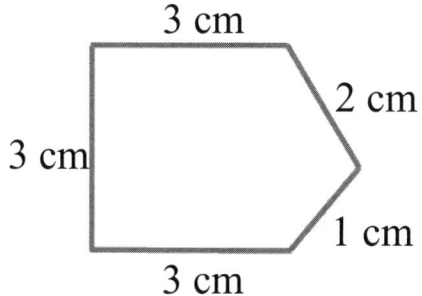

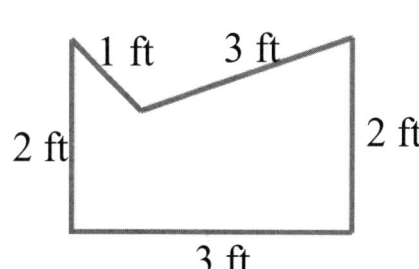

5. _____   6. _____

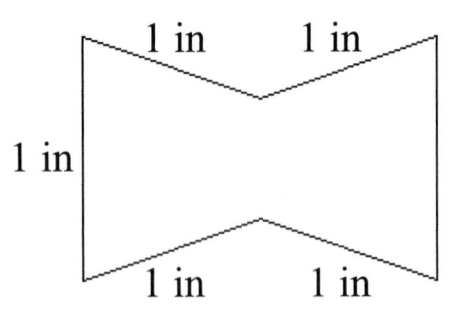
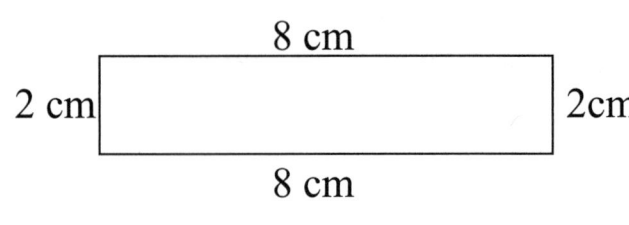

7. _____   8. _____

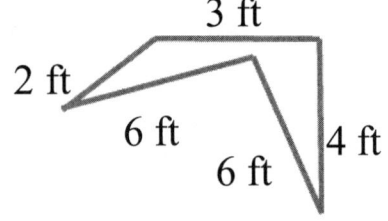
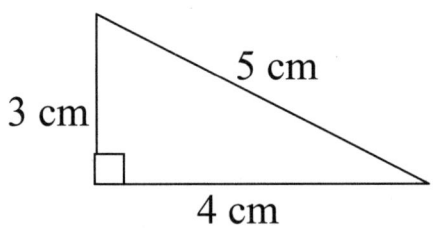

9. _____   10. _____

11. How far would you walk if you went around a block that measures 3 kilometers on each side?

12. What is the perimeter of a rectangle that measures 4 inches by 10 inches?

13. If a square measures 10 centimeters on one side, what is the square's perimeter?

14. Abdul has a rectangular garden 3 feet long and 2 feet wide. How long is the fence that goes around his garden?

Definition: *Area* is the number of square units in a figure.

1 square unit   4 square units   10 square units

Now try these:

Find the area of each figure.

1. _____ square units   2. _____ square units

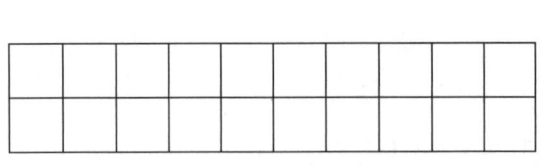

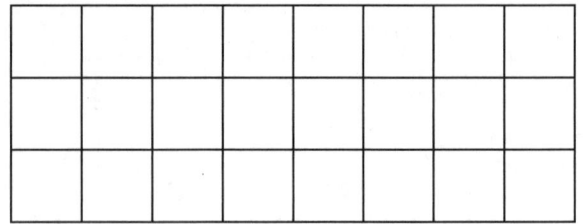

3. _____ square units   4. _____ square units

To find the area of a figure, multiply the length by the width.

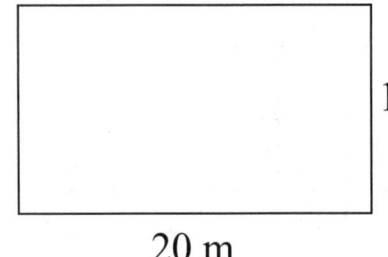
10 m

20 meters × 10 meters

20 × 10 = 200 square meters

20 m

Now try these:

Find the area of each figure.

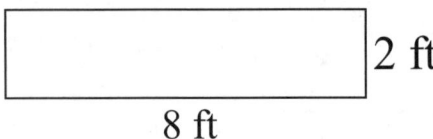
2 ft

8 ft

3 cm

13 cm

1. _____ square feet   2. _____ square cm

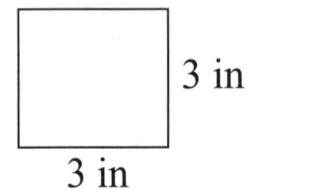
3 in

3 in

4 m

9 m

3. _____ square inches   4. _____ square m

5. Hans was making a fence around his yard for his dog. How many square feet of space will be needed if the fence is 7 feet long and 9 feet wide?

6. A square measures 10 centimeters on one side, what is the area?

7. Morgan made a garden that is 12 feet long and 6 feet wide. What is the area of her garden?

8. A book is 8 inches by 10 inches. What is the surface area of the book?

Cluster III

Mathematical Processes – Measurement

Units of Measure:

Customary/Traditional System

Length	inch (in.) foot (ft.) yard (yd.) mile (mi.)
Weight	ounce (oz.) pound (lb.) ton (t)
Capacity	cup (c) pint (pt.) quart (qt.) gallon (gal.) teaspoon (tsp.) tablespoon (Tbsp)
Area	square units square inches (sq.in.) square feet (sq.ft.) square miles (sq.mi.)
Volume	cubic units cubic inches (cu.in.) cubic feet (cu.ft.)
Temperature	degrees Fahrenheit
Time	second (sec.) minute (min.) hour (hr.)

Metric System

Length	millimeter (mm) centimeter (cm) meter (m) kilometer (km)
Weight	gram (g) kilogram (kg)
Capacity	liter (l)
Area	square centimeters (sq.cm)
Temperature	degrees Celsius

Measurement Conversion

1 foot	= 12 inches	
1 yard	= 3 feet	= 36 inches
1 mile	= 5,280 feet	
1 cup	= 8 ounces	
1 pint	= 2 cups	= 16 ounces
1 quart	= 2 pints	= 32 ounces
1 gallon	= 4 quarts	= 128 ounces

Calendar

January

S	M	T	W	T	F	S
				1	2	3
4	5	6	7	8	9	10
11	12	13	14	15	16	17
18	19	20	21	22	23	24
25	26	27	28	29	30	31

February

S	M	T	W	T	F	S
1	2	3	4	5	6	7
8	9	10	11	12	13	14
15	16	17	18	19	20	21
22	23	24	25	26	27	28

March

S	M	T	W	T	F	S
	1	2	3	4	5	6
7	8	9	10	11	12	13
14	15	16	17	18	19	20
21	22	23	24	25	26	27
28	29	30	31			

April

S	M	T	W	T	F	S
				1	2	3
4	5	6	7	8	9	10
11	12	13	14	15	16	17
18	19	20	21	22	23	24
25	26	27	28	29	30	

May

S	M	T	W	T	F	S
						1
2	3	4	5	6	7	8
9	10	11	12	13	14	15
16	17	18	19	20	21	22
23	24	25	26	27	28	29
30	31					

June

S	M	T	W	T	F	S
		1	2	3	4	5
6	7	8	9	10	11	12
13	14	15	16	17	18	19
20	21	22	23	24	25	26
27	28	29	30			

July

S	M	T	W	T	F	S
				1	2	3
4	5	6	7	8	9	10
11	12	13	14	15	16	17
18	19	20	21	22	23	24
25	26	27	28	29	30	31

August

S	M	T	W	T	F	S
1	2	3	4	5	6	7
8	9	10	11	12	13	14
15	16	17	18	19	20	21
22	23	24	25	26	27	28
29	30	31				

September

S	M	T	W	T	F	S
			1	2	3	4
5	6	7	8	9	10	11
12	13	14	15	16	17	18
19	20	21	22	23	24	25
26	27	28	29	30		

October

S	M	T	W	T	F	S
					1	2
3	4	5	6	7	8	9
10	11	12	13	14	15	16
17	18	19	20	21	22	23
24	25	26	27	28	29	30
31						

November

S	M	T	W	T	F	S
	1	2	3	4	5	6
7	8	9	10	11	12	13
14	15	16	17	18	19	20
21	22	23	24	25	26	27
28	29	30				

December

S	M	T	W	T	F	S
		1	2	3	4	
5	6	7	8	9	10	11
12	13	14	15	16	17	18
19	20	21	22	23	24	25
26	27	28	29	30	31	

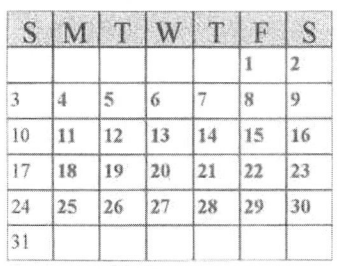

1. What months come after September?

2. What months come before July?

3. What is the day before January 1?

4. What is the next date after October 31?

5. What is the third month?

6. What is the shortest month?

Measuring in Inches

| | 1 | 2 | 3 | 4 | 5 | 6 |

Use your ruler to measure the length of each drawing.

_____ inches

_____ inches

_____ inches

_____ inches

_____ inches

_____ inches

Measuring in Centimeters

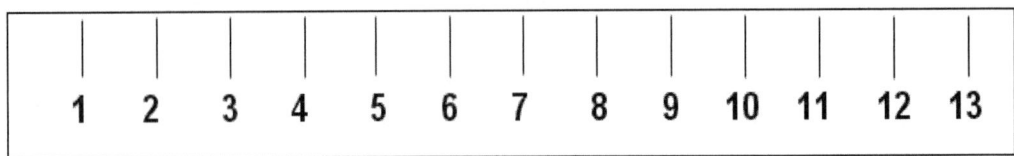

Use a centimeter ruler to measure each drawing.

_____ centimeters

_____ centimeters

_____ centimeters

_____ centimeters

_____ centimeters

_____ centimeters

_____ centimeters

Guess and Measure

Use your centimeter ruler (cm). Guess how long each line is. Measure and see how accurately you guessed.

Guess: _____ centimeters

Measure: _____ centimeters

Guess: _____ centimeters

Measure: _____ centimeters

Guess: _____ centimeters

Measure: _____ centimeters

Guess: _____ centimeters

Measure: _____ centimeters

Estimating Foot (ft), Yard (yd), and Mile (mi)

Circle the best estimate for the length of each.

1. height of a basketball hoop   10 ft   10 yd   10 mi

2. length of a football field   100 ft   100 yd   100 mi

3. height of a woman   5 yd   5 mi   5 ft

4. length of a highway   60 ft   60 mi   60 yd

5. long distance to drive   300 ft   300 yd   300 mi

6. your height   4 ft   4 yd   4 mi

7. distance of a long race   1 yd   1 mi   1 ft

To measure heavier objects, we use pounds.

1 pound = 16 ounces

A bag of 4 apples weighs about 1 pound.

Write ounces (oz) or pounds (lb) next to each item.

1. an apple _____

2. a dog _____

3. a table _____

4. an orange _____

5. a watermelon _____

6. a television _____

7. a book _____

8. a tooth _____

9. a bed _____

10. a sock _____

11. a pencil _____

12. a turkey _____

Estimating Ounces (oz), Pounds (lb), and Tons (T)

Circle the best unit of measure for each.

1. envelope   oz   lb   T

2. car   oz   lb   T

3. airplane   oz   lb   T

4. feather   oz   lb   T

5. peach   oz   lb   T

6. desk   oz   lb   T

7. cat   oz   lb   T

8. box of juice   oz   lb   T

9. Tim and his father made hamburgers for the football team. About how much meat did they need?

    8 oz   8 lb   8 T

10. Gary went to the store for potatoes to make potato pancakes for his family. About how many potatoes did he need?

    5 lbs   5 oz   5 T

11. Clarissa bought 2 bags of flour to make cookies for her birthday party. About how much did the bags weigh?

    10 oz   10 T   10 lbs

12. Ms. Patel had 10 rubber bands on her desk. About how much did they weigh?

    10 oz   10 T   10 lbs

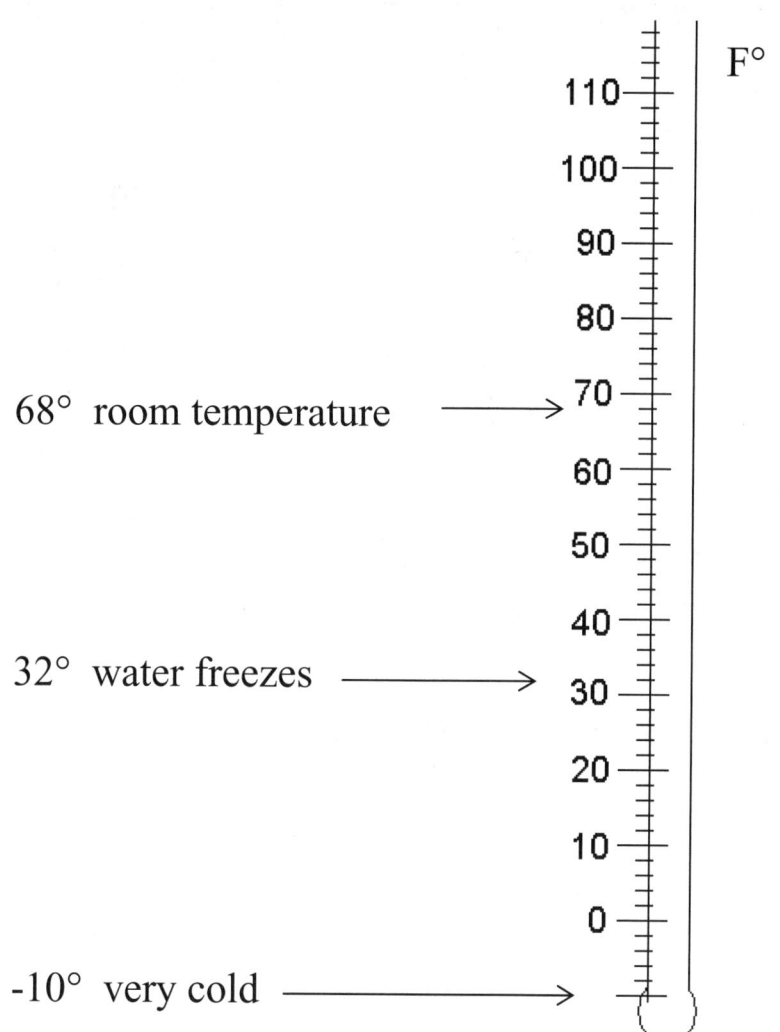

68° room temperature

32° water freezes

-10° very cold

Circle the most reasonable estimate.

1.	bike riding weather	76°F	32°F	10°F
2.	skiing weather	95°F	62°F	30°F
3.	swimming weather	82°F	64°F	56°F
4.	water freezes	68°F	32°F	47°F

Write the temperature.

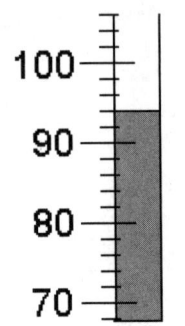   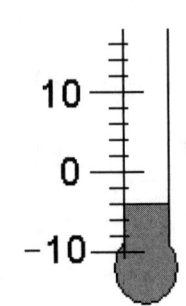   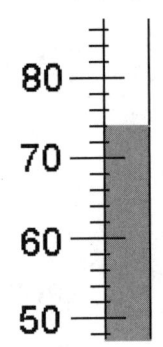

5. _____°F   6. _____°F   7. _____°F

8.   Which temperature is colder: -5°F or -10°F?

Solid Figures

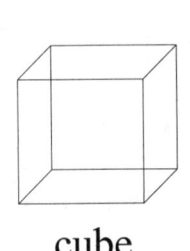   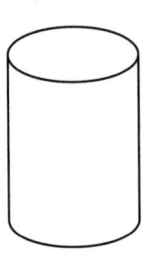   

    cube   cylinder   cone

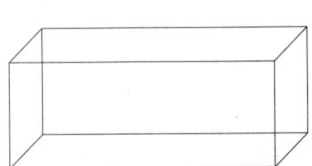   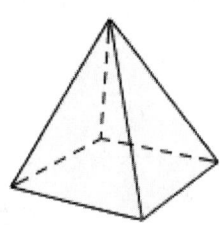   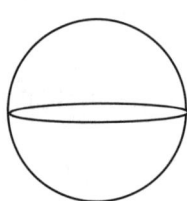

rectangular solid   pyramid   sphere

Which object is an example of a cylinder?

 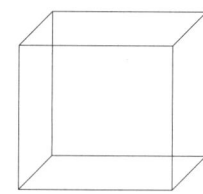 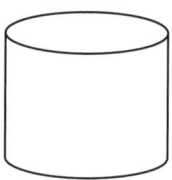 

Liquid Measure

1 pint = 2 cups
1 quart = 2 pints or 4 cups
1 gallon = 4 quarts, 8 pints, or 16 cups

1. How many cups of milk are in 2 quarts?

2. Mr. Jackson's class is having a party. There are 26 students in his class. If each child drinks 1 cup of punch, how many gallon containers of punch will Mr. Jackson have to buy? Show and explain your answer.

3. How many quarts equal 3 gallons of juice?

4. Which is a greater amount of milk: 1 gallon or 32 cups? Explain your answer.

5. Which is less: 16 cups of water or 2 pints? Explain your answer.

Telling Time

What time is it?

Answer: 1:00

Answer: 5:00

The smaller arrow always points to the hour. The larger arrow always points to the minutes.

When telling time, you can skip count by 5 starting with 1 and ending with 12: 5, 10, 15, 20, 25, 30, 35, 40, 45, 50, 55, 60. Remember: Every 60 minutes starts a new hour.

Now try these:

1.   What number is the hour hand pointing to?

A.   12

B.   6

C.   5

D.   3

2. Fill in the rest of the clock with the correct numerals.

3. Which clock shows that it is 2:00?

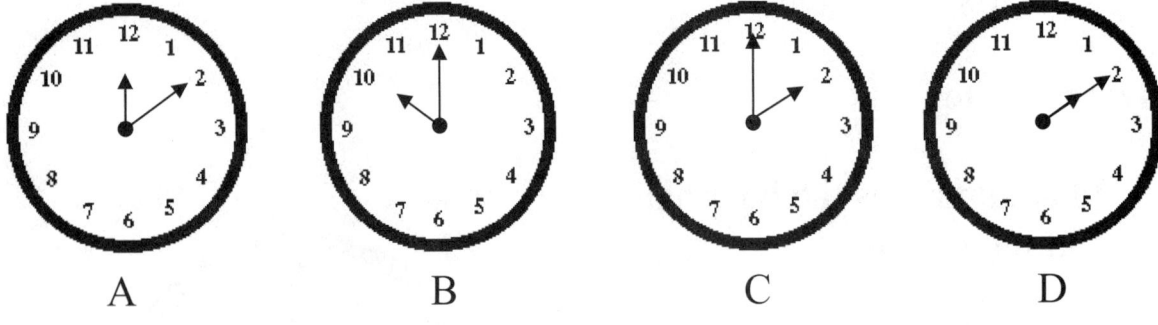

A   B   C   D

4. Which clock shows that it is 4:30?

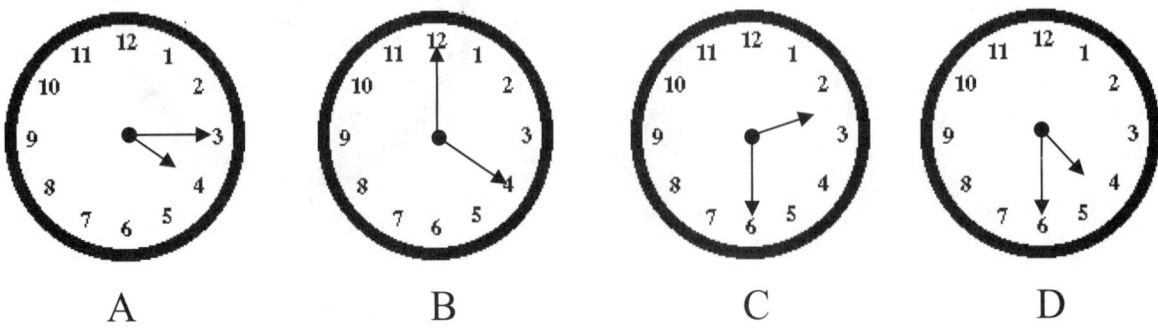

A   B   C   D

5. Draw the hands to show 5:30.

6. Draw the hands to show 8:00.

Estimating Time

Time is measured in different units.

60 seconds = 1 minute
60 minutes = 1 hour
24 hours = 1 day
365 days = 1 year

Estimate the time it will take:

1. to tie your sneakers   1 minute   1 hour   1 day

2. to mow the lawn   1 minute   2 hours   4 days

3. to brush your teeth   1 minute   1 hour   1 day

4. to do your homework   1 minute   2 hours   4 days

5. to wash your hands   1 minute   1 hour   1 day

6. Karen's piano practice begins at 4:00 P.M. and ends at 6:15 P.M. How long is Karen's piano practice?

 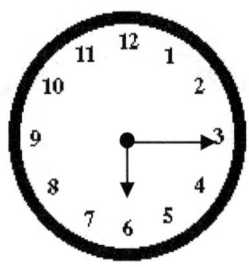

7. Mark spent from 5:30 A.M. to 2:15 P.M. studying for a test. How long did he study?

8. Robin starts her homework at 6:30 P.M. Her homework takes 2 hours and 20 minutes. When will Robin finish her homework?

9. Drew leaves home at 8:35 A.M. to walk to school. It takes him 15 minutes. What time will Drew arrive at school?

10. Math class for third grade begins at 8:45 A.M. and lasts for 1 hour and 20 minutes. When will math class end?

11. An after-school computer class at Martin Luther King School begins at 3:20 P.M. and ends at 5:00 P.M. How long is the computer class?

12. Kelly's basketball practice begins at 6:00 P.M. It is one hour and thirty minutes long. What time will Kelly finish her practice?

13. Michelle arrived at a party at the time shown on clock 1. She left at the time shown on clock 2. How long was she at the party?

Clock 1

2:10 PM

Clock 2

7:17 PM

14. A soccer game began at 2:00 P.M. The team played two 45-minute halves. What time did the game end?

15. A movie started at 7:00 P.M. and ended at 8:45 P.M. How long was the movie?

16. Vicki works at a bank. She begins work at 8:00 A.M. and ends at 4:30 P.M. Vicki has 30 minutes for lunch. How many hours does Vicki work?

17. At 4:30 P.M. Matthew got up from a nap. He slept for 2 hours and 45 minutes. What time did he go to bed?

18. Sue helps at the library in school. She helps 1 hour before school and 1 hour after school. She did this for 5 days. How many hours did she help the librarian?

19. Ben works at the New Jersey State Aquarium. These are the hours he works in a week. On Monday he works 8 hours, Tuesday 6 hours, Wednesday 6 hours, Thursday off, Friday off, Saturday 8 hours and Sunday 6 hours. How many hours did Ben work this week?

20. Grandpa is cooking a turkey. He put the turkey in the oven at 8:00 A.M. It takes 4 hours and 45 minutes for the turkey to cook. The turkey must cool for 30 minutes before Grandpa can cut it. What time can the turkey dinner be served?

21. Jose left school at 2:55 P.M. He arrived home 15 minutes later. What time did he arrive home? Draw the hands on the clock to show the time.

22. Which clock shows the same time as the digital clock?

5:20

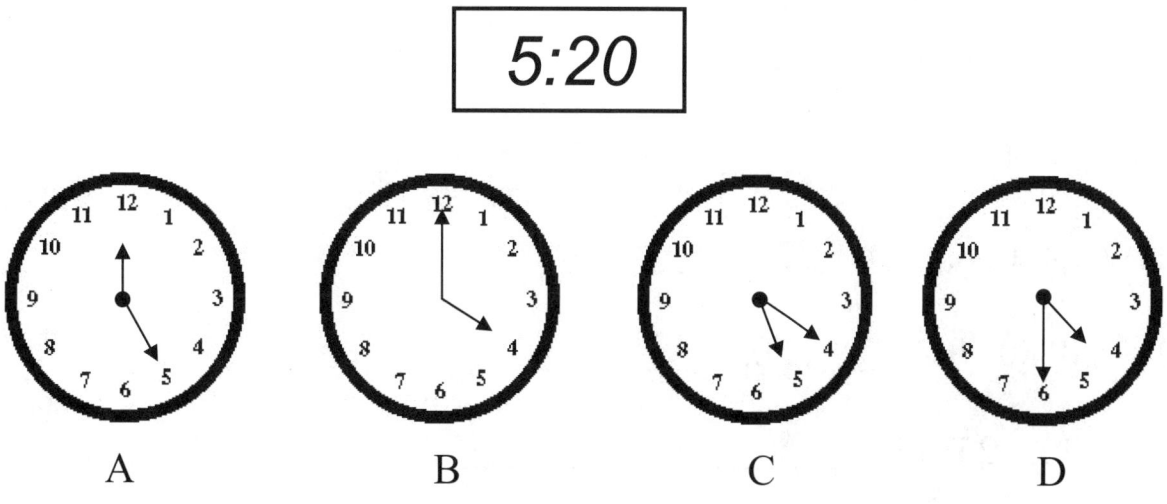

A   B   C   D

Cluster IV

Patterns and Algebra

Patterns can be used to predict what comes next.

What number is missing from the pattern?

> 3, 5, 7, 9, ___, 13, 15

Look at the number sequence, and ask yourself:

- Are the numbers getting larger or smaller?
- What was done to the 3 to get to the 5?
- What was done to the 5 to get to the 7?
- What was done to the 7 to get to the 9?
- What was done to the 13 to get to the 15?

```
3    5    7    9    ___    13    15
 V    V    V    V          V     V
+2   +2   +2   +2         +2    +2
```

The pattern is to add 2 to continue the pattern.

Find the pattern.

1. 7, 9, 11, 13, _____, _____

2. 2, 4, 6, 8, _____, 12, _____

3. 90, 80, 70, _____, 50, _____

4. 120, 125, 130, _____, 140, _____

5. 990, 992, 994, 996, 998, _____

6. red, blue, green, green, red, _____, green, _____

7. □, □, ○, ☆, ○, □, □, ○, ☆, ○, □, _____, _____

8. ◆, X, ○, □, □, ◆, X, ○, □, □, _____, _____, _____

9. 4, 8, 12, _____, 20, _____, 28, 32

10. 3, 2, 4, 3, 6, _____, _____, _____

11. What two numbers complete this pattern?

    27, 24, 21, _____, _____, 12, 9

    Describe the pattern.

12. Look at the numbers below. What will be the next number?

    1, 2, 4, 7, _____

13. What comes next?

    3, 6, 6, 12, 12, _____

    A. 24, 26   C. 21, 21,

    B. 18, 24   D. 24, 24

14. ☐ is to | as ⬚ is to _____.

A. ╱

C. ╱ (dashed)

B. ✛

D. ┊

15. What are the next two numbers in the pattern?

634, 630, 626, _____, _____

16. At the diner a small soda costs 86¢. A medium soda costs 92¢. A large soda costs 98¢. If the pattern continues, how much will an extra-large soda cost?

17. On vacation at the beach you find 5 seashells the first day, 6 on the second day, and 8 on the third day. If the pattern continues, how many seashells will you find on the fifth day?

What is the total number of seashells that you have found on the five days?

18. Connor played 4 soccer games in March, 9 soccer games in April, and 14 soccer games in May. If the pattern continues, how many games will Connor play in June?

How many soccer games did Connor play during those four months?

Function Tables

Complete the following tables.

1. Rule: Add 10	
Input	Output
6	
20	
	73
91	
	41

2. Rule: Add 148	
Input	Output
112	
	180
567	
	202
	424

3. Rule: Subtract 15	
Input	Output
68	
58	
	39
62	
20	

4. Rule: Subtract 121	
Input	Output
286	
	224
544	
	808
687	

5. Rule: Multiply by 4	
Input	Output
3	
6	
8	
	36

6. Rule: Multiply by 6	
Input	Output
6	
	60
9	
3	

Completing Number Sequences

$$8 + 2 = 10$$

↑   ↑   ↑

addend addend   sum

Find the missing addend.

Example: $6 + ? = 8$

Ask yourself: What number plus 6 equals 8? Think: $6 + 1 = 7, 6 + 2 = 8$
(Stop here because $6 + 2 = 8$.)
Conclusion: In the example, if $6 + ? = 8$, then $? = 2$.

Critical thinking: $6 + 2 = 8$ and $8 - 2 = 6$ and $8 - 6 = 2$. Do you see another way of solving this problem?

Example: $3 + ? = 15$

Think:
$3 + 1 = 4$	$3 + 2 = 5$	$3 + 3 = 6$	$3 + 4 = 7$
$3 + 5 = 8$	$3 + 6 = 9$	$3 + 7 = 10$	$3 + 8 = 11$
$3 + 9 = 12$	$3 + 10 = 13$	$3 + 11 = 14$	$3 + 12 = 15$

or $15 - 3 = ?$
$12 = ?$

Now try these:

1. $6 + ? = 8$ $8 - \underline{\hspace{2cm}} = ?$ $? = \underline{\hspace{3cm}}$

2. $5 + ? = 15$ $15 - \underline{\hspace{2cm}} = ?$ $? = \underline{\hspace{3cm}}$

3. $9 + ? = 20$ $20 - \underline{\hspace{2cm}} = ?$ $? = \underline{\hspace{3cm}}$

4. $2 + ? = 18$ $18 - \underline{\hspace{2cm}} = ?$ $? = \underline{\hspace{3cm}}$

Critical Thinking

Andrea spent a day at the dog show. Andrea saw that there were a total of 15 dogs at the show. 7 dogs did tricks in the morning. How many dogs performed tricks in the afternoon?

Think:	7	+	?	=	15
	number of dogs that did tricks in the morning		number of dogs that did tricks in the afternoon		total number of dogs that did tricks that day

<table>
<tr><td>Method 1</td><td>or</td><td>Method 2</td></tr>
<tr><td>7 + 1 = 8</td><td></td><td>7 + ? = 15</td></tr>
<tr><td>7 + 2 = 9</td><td></td><td></td></tr>
<tr><td>7 + 3 = 10</td><td></td><td>means</td></tr>
<tr><td>7 + 4 = 11</td><td></td><td></td></tr>
<tr><td>7 + 5 = 12</td><td></td><td>15 − 7 = ?</td></tr>
<tr><td>7 + 6 = 13</td><td></td><td>8 = ?</td></tr>
<tr><td>7 + 7 = 14</td><td></td><td></td></tr>
<tr><td>7 + 8 = 15</td><td></td><td></td></tr>
</table>

Answer: Andrea saw 8 dogs do tricks in the afternoon.

Now try these:

1. Last year the Allwags won 8 games. This year they won 20 games. How many more games did they win this year?

2. George's teacher gave 30 problems to do for homework during spring break last year. This year George's teacher gave 50 problems to do for homework during spring break. How many more problems did he give this year?

3. Courtney read 6 books last year. This year she read 14 books. How many more books did she read this year?

4. Daniel had 10 baseball cards on Monday. On Thursday he had 18 baseball cards. How many more baseball cards did he have on Thursday than on Monday?

5. Sammy saw 5 volleyball teams playing. Each had 8 players, for a total of 40 players. Which number sentence is NOT a method for expressing this fact?

   A. $40 - 8 - 5 = 27$

   B. $5 \times 8 = 40$

   C. $8 \times 5 = 40$

   D. $40 \div 8 = 5$

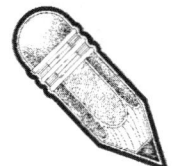

Cluster V

Data Analysis, Probability, and Discrete Mathematics

<u>Chances (Probability)</u>

The chances (probability) of an event tells us the likelihood that an event will occur. Probability is written as a number between 0 and 1. A probability of 0 means that the event will not happen, while a probability of 1 means that the event will happen.

$$\text{Probability of an Event} = \frac{\text{Number of favorable outcomes}}{\text{Total number of possible outcomes}}$$

Example 1.　There are 7 blue and 3 yellow marbles in a bag. What is the probability of picking a red marble from the bag?

$$P(E) = \frac{\text{Number of favorable outcomes}}{\text{Total number of possible outcomes}}$$

$$P(red) = \frac{0}{7+3} = \frac{0}{10} = 0$$

Note: A probability of 0 means that the event will not occur.

Example 2.　In example 1, what is the probability of picking a blue marble?

$$P(E) = \frac{\text{Number of favorable outcomes}}{\text{Total number of possible outcomes}}$$

$$P(blue) = \frac{7}{7+3} = \frac{7}{10}$$

Note: A probability between 0 and 1 means the event is possible.

Example 3.   There are 10 purple marbles in a bag. What is the probability of picking a purple marble from the bag?

$$P(E) = \frac{\text{Number of favorable outcomes}}{\text{Total number of possible outcomes}}$$

$$P(\text{purple}) = \frac{10}{10} = 1$$

Note: A probability of 1 means that the event will occur.

Example 4.   What are the chances (probability) that the pointer will stop on number 1?

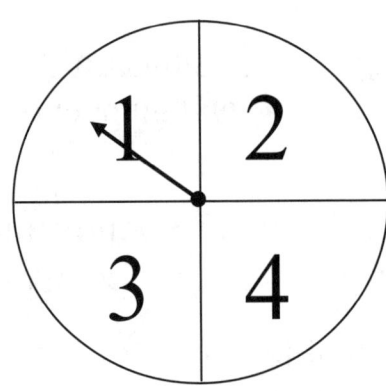

$$P(E) = \frac{\text{Number of ones}}{\text{Total number of numbers}} = \frac{1}{4}$$

What are the chances (probability) that the pointer will stop on number 2?

$$P(E) = \frac{\text{Number of twos}}{\text{Total number of numbers}} = \frac{1}{4}$$

What are the chances (probability) that the pointer will stop on number 3?

$$P(E) = \frac{\text{Number of threes}}{\text{Total number of numbers}} = \frac{1}{4}$$

In example 4, which number do you have the best chance (probability) of spinning? Why?

_____

_____

_____

_____

Example 5.   A number cube with the numbers 1, 2, 3, 4, 5, and 6 (one number on each of its six faces) is rolled once. What is the chance (probability) of getting a 2?

$$P(2) = \frac{\text{Number of ways to roll a 2}}{\text{Total number of numbers}} = \frac{1}{6}$$

Now try these:

1.   What are the chances (probability) of picking the number 1 from the group of numbers below without looking (i.e., at random)?

1	2	3
	1	
1	2	1

$$P(E) = \frac{\text{Number of ones}}{\text{Total number of numbers}} = \frac{4}{7}$$

133

2. Jerry has 10 pairs of socks in his drawer. Three of those pairs are white. What are the chances (probability) of picking a pair of white socks without looking?

$$P(E) = \frac{\text{Number of white pairs of socks}}{\text{Total number of pairs of socks}} = \frac{3)}{10}$$

3. Make a set of 10 number cards. Write the number 4 on three cards, 2 on four cards, 5 on one card, and 7 on two cards. Answer the following questions.

a. How many cards have the number 4 on them?   3

b. How many cards have the number 2 on them?   4

c. How many cards have the number 5 on them?   1

d. How many cards have the number 7 on them?   2

e. What are the chances (probability) of picking a card with a 4 on it without looking? $\frac{3}{16}$

f. What are the chances (probability) of picking a card with a 2 on it without looking? $\frac{4}{16}$

g. What are the chances (probability) of picking a card with a 5 on it without looking? $\frac{1}{10}$

h. What are the chances (probability) of picking a card with a 7 on it without looking? $\frac{2}{10}$

4. Mr. Peterson uses a spinner to choose the activity for his physical education class. He will spin one of the spinners to decide what activity his class will do.

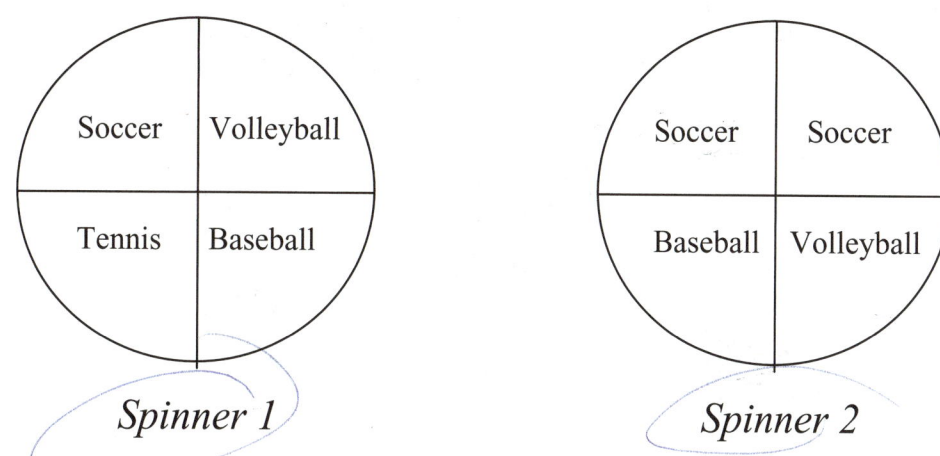

a. What are the chances (probability) of landing on soccer using spinner 1?

b. What are the chances (probability) of landing on soccer using spinner 2?

c. What spinner should Mr. Peterson use if he wants to play soccer?

Show your work and explain your answer.

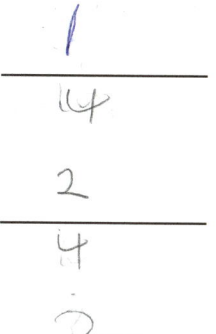

5. Stephanie is going to toss a ball into one of the boxes that makes up the grid shown below.

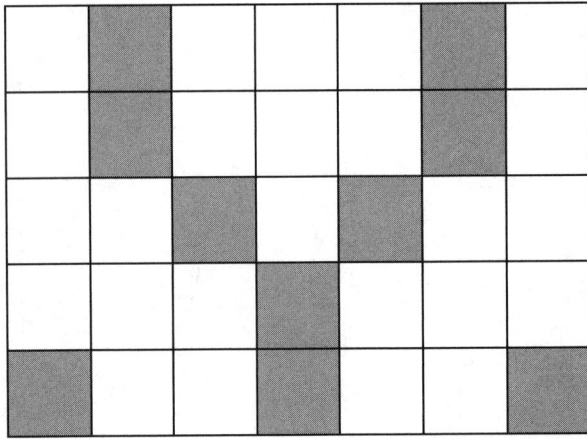

a. How many total boxes are there? _____

b. How many shaded boxes are there? _____

c. How many boxes are not shaded? _____

d. What are the chances (probability) that the ball will land in a shaded box? _____

e. What are the chances (probability) that the ball will land in a box that is not shaded? _____

6. A number cube with the numbers 1, 2, 3, 4, 5, and 6 (one number on each of its 6 faces) is rolled once. What are the chances (probability) of getting the following?

a. P(1) =   _____

b. P(2) =   _____

c. P(3) =   _____

d. P(4) =   _____

e. P(5) =   _____

f. P(6) =   _____

g. P(1 or 2) = P(1) + P(2) =   _____

h. P(1, 2, or 3) = P(1) + P(2) + P(3) =   _____

i. P(a number greater than 4) = P(5) + P(6) =   _____

j. P(a number less than 4) = P(1) + P(2) + P(3) =   _____

Counting Principle

Example: The cafeteria is offering one sandwich and one drink for lunch.

<u>Sandwiches</u>   <u>Drinks</u>
Bologna   Milk
Turkey   Juice
Cheese

How many combinations of one sandwich and one drink are there?

Solution: <u>Step 1</u>   <u>Step 2</u>
              There are 3 sandwiches   After you choose a sandwich
              to choose from   there are 2 drinks to choose from

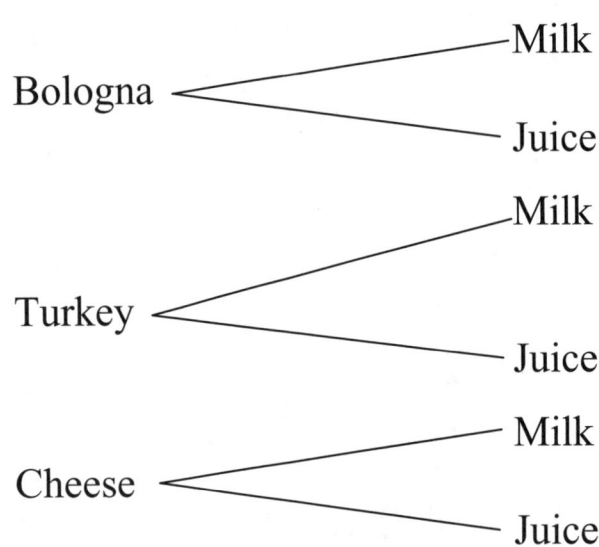

(Note: The above diagram is called a tree diagram.)

Answer: There are 6 difference combinations.

1. Bologna and milk   4. Turkey and juice
2. Bologna and juice   5. Cheese and milk
3. Turkey and milk   6. Cheese and juice

Now try these:

1. The ice cream store offers 4 flavors (vanilla, chocolate, strawberry and butter pecan) and 2 cones (regular and sugar).

    a. Complete the tree diagram below.

    _____   _____
                            <
                               _____

    _____   _____
                            <
                               _____

    _____   _____
                            <
                               _____

    _____   _____
                            <
                               _____

    b. How many different combinations are there?   _____

    c. List all the different combinations below.

    _____ and _____, _____ and _____,

    _____ and _____, _____ and _____,

    _____ and _____, _____ and _____,

    _____ and _____, _____ and _____.

2. The students in a 3rd grade class are making a banner for their class trip. They can choose from 3 types of material and 5 colors. The choices are as shown.

<u>Materials</u>
Cotton
Rayon
Wool

<u>Colors</u>
Red   Green
Blue   Yellow
Pink

a. Complete the tree diagram below.

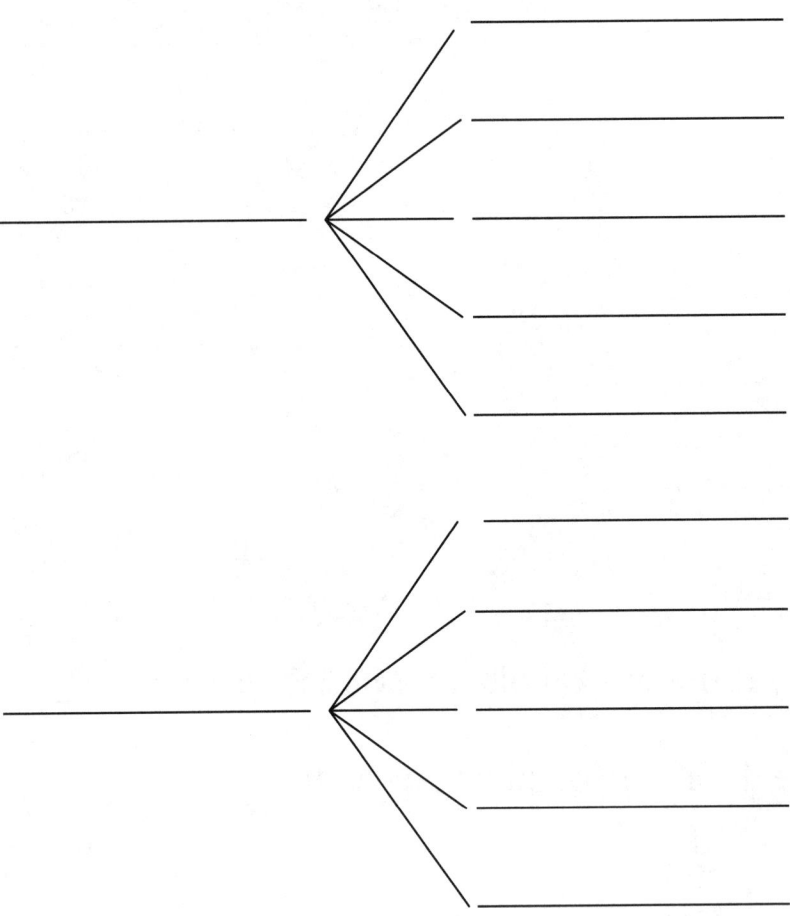

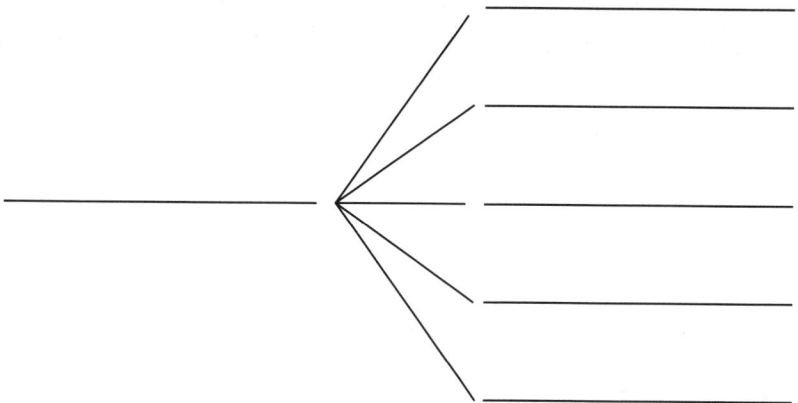

b. How many different combinations are there?_____

c. List all the different combinations.

_____ and _____, _____ and _____,

_____ and _____, _____ and _____,

_____ and _____, _____ and _____,

_____ and _____, _____ and _____,

_____ and _____, _____ and _____,

_____ and _____, _____ and _____,

_____ and _____, _____ and _____,

_____ and _____

Multiplication Principle:

Definition: When you choose one item from one set of items and you choose a second item from a different set of items, you can determine the number of different combinations by multiplying the number of items in set one by the number of items in set two.

Example: The cafeteria is offering the following items for lunch. You must choose one item from set 1 and 1 item from set 2.

Set 1	Set 2
Pizza	Milk
Tacos	Juice
Sandwich	

There are 3 items in set 1 and 2 items in set 2. Thus, the total number of different combinations is: **3 × 2 = 6**.

Answer: There are 6 combinations to choose from.

Now try these:

1.   Michael wants to buy a cheese pizza. He can choose one type of crust and one extra topping from the choices below. How many different choices does Michael have?

Crust	Topping
Regular	Extra Cheese
Thin	Pepperoni
	Sausage

Answer: _____

2.   Candace is learning how to bowl. Her mother took her to a store to buy her a bowling ball and a pair of bowling shoes. How many different choices does Candace have?

Bowling Ball Colors	Shoe Colors
Black	White Tan
Pink	Tan Purple
Red	Red

Answer: _____

3. Ted has 2 shirts (black and white) and 4 ties (green, red, blue, and yellow). How many different outfits can he make if the outfit consists of 1 shirt and 1 tie?

   Answer: _____

4. Subs R Us has a special for lunch. A special is a sandwich and a drink. There are 3 sandwiches and 3 drinks. How many lunch specials can you select?

   Subs R Us Specials

Sandwiches	Drinks
Ham	Soda
Turkey	Milk
Roast Beef	Juice

   Answer: _____

5. How many different outfits can be made with 5 hats and 3 shirts?

   Answer: _____

6. How many different outfits can be made with 7 shirts and 2 skirts?

   Answer: _____

7. Rose wants to arrange her dolls and stuffed animals on her bed. If she has 4 dolls and 3 stuffed animals, how many different arrangements can she make?

   Answer: _____

A More Difficult Problem

Example 1: Cameron, Joe, and Gordon want to shake hands with each other only once. How many distinct handshakes are there?

Solution:

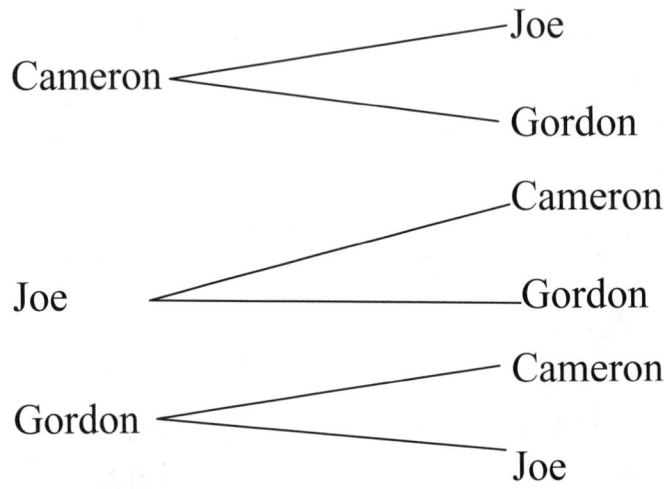

List all the handshakes.

1. Cameron and Joe
2. Cameron and Gordon
3. Joe and Cameron
4. Joe and Gordon
5. Gordon and Cameron
6. Gordon and Joe

Notice that some of the handshakes have the same people.

1. Cameron and Joe } same people } counts as 1
3. Joe and Cameron } shaking hands } distinct handshake

2. Cameron and Gordon } same people } counts as 1
5. Gordon and Cameron } shaking hands } distinct handshake

4. Joe and Gordon } same people } counts as 1
6. Gordon and Joe } shaking hands } distinct handshake

When you count the distinct handshakes, you get 3.

Combinations:

Rule: When choosing two members from the same set, multiply the original number times one less than the original number and divide that product by 2. Thus, you will eliminate the duplicate counts.

Note: The original number of people was 3 (Cameron, Joe, and Gordon). One less than 3 equals 2.

$$3 \times 2 = 6$$
$$6 \div 2 = 3$$

Conclusion: There were 3 distinct handshakes.

Example 2: Brianna has 4 textbooks in her desk. She wants to bring home two textbooks tonight. How many different pairs of textbooks could she take home? (Be careful of duplication.)

Step 1. Number of books to pick from to choose the first book _____

Step 2. Number of books to pick from after the first book is chosen _____

Step 3. _____ × _____ = _____

Number of books to pick for the first choice   Number of books to pick from for the second choice   Product including duplication

Step 4. _____ ÷ 2 = _____

Product including duplication   Total number of different pairs of textbooks

Range, Median, Mode, and Mean

Range: The difference between the greatest and least numbers.

Find the greatest number in the list and the least number in the list. Subtract the 2 numbers and you will find the range.

Example: Given the set of data: {3, 4, 2, 5, 7, 4}, what is the range? 7 is the greatest number; 2 is the least number.

$$7 - 2 = 5$$
So 5 is the range.

Median: The middle number when the data is listed in order.

Rewrite the list of numbers in order from least to greatest. Mark off one at a time from each end to find the median.

Example: 1, 2, 3, 4, 4, 4, 5, 5, 6
1, 2, 3, 4, (4,) 4, 5, 5, 6

So 4 is the median.

Mode: The number that occurs most often.

Look at the ordered list and find the number written most often.

Example: 1, 2, 3, 4, 4, 4, 5, 5, 6
4 occurs more times, so 4 is the mode.

Mean: The average of a set of data. Add all the numbers in the set of data then divide by the number of numbers.

Example: What is the mean for: 2, 2, 3, 4, and 4.

$$\frac{2+2+3+4+4}{5}$$

$$\frac{15}{5} = 3$$

1. Given the set of data, {6, 4, 2}:

   a. Which is the greatest number in this group? _____

   b. Which is the least number? _____

   c. What is the difference between the greatest and
      the least number (Range)? _____

   d. What is the median of this group of numbers? _____

   e. What is the mean? _____

2. Given the set of data {8, 7, 6, 6, 1}:

   a. Which number is the median of this group of numbers?_____

   b. What is the range of this group of numbers? _____

   c. What is the mode of this group of numbers? _____

Favorite Sports	
Sport	Votes
Baseball	12
Hockey	7
Basketball	12
Track	10
Football	9

3. a. What is the median? _____

b. What is the range? _____

c. What is the mode? _____

d. What is the mean? _____

Draw a picture representation of the mean to show your answer.

Number of Pet Owners in Mrs. Jones' 4th Grade Class

```
Number of People
      X
      X     X
      X     X     X
      X     X     X           X
      X     X     X     X     X     X
      0     1     2     3     4     5
              Number of Pets
```

4. a. How many people have 2 pets? _____

148

b.   Does everyone have a pet? Explain your answer.

c.   What is the highest number of pets a person has?   _____

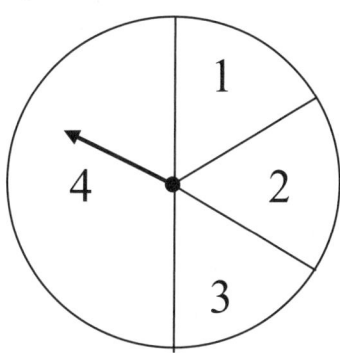

5.   What number is the most likely outcome of the spinner? Explain your answer.

_____

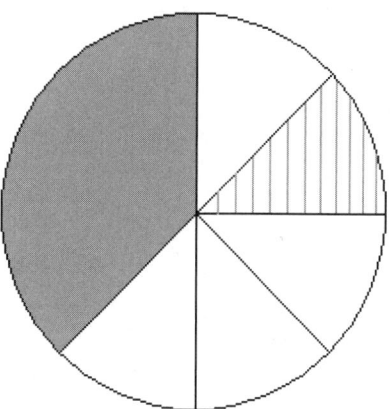

6.   What is the most likely outcome of the spinner?

_____

Use this chart for question #7.

	Number of Soccer Goals						
		X					
		X			X		
	X	X			X		
	X	X	X		X		X
	0	1	2	3	4	5	6

P
L
A
Y
E
R
S

GOALS

7. a. How many players had 4 goals? _____

    b. How many players scored less than 3 goals? _____

    c. What is the range of data? _____

    d. What is the median? _____

    e. What is the mode? _____

Use this spinner for question #8.

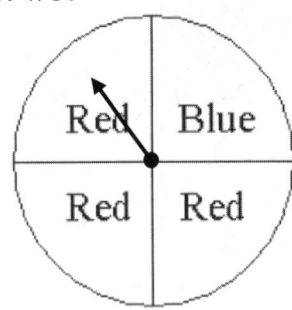

8. a. Carol decides to spin the pointer. On what color is the pointer most likely to stop? Explain your answer.

b. Carol's math teacher says, "The pointer has a 0 chance of stopping on yellow." What do you think Carol's teacher means?

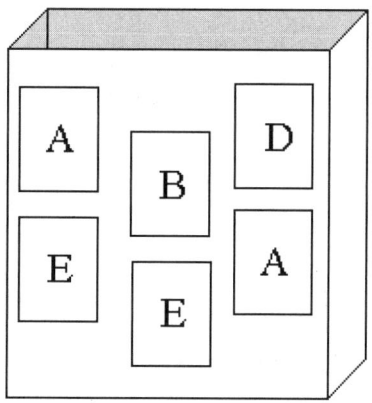

9. a. Suppose you pick out a letter from the box. What is the probability that it will be an "A"? _____

b. What is the probability that it will be a "D"? _____

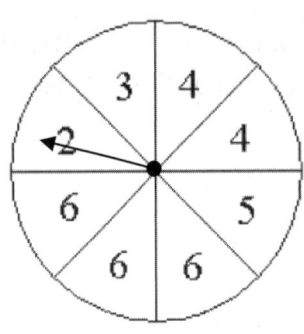

10. Suppose you spin one time. On which number is the spinner most likely to land?

A. 3   C. 4

B. 2   D. 6

11. Two students are playing a game. Player 1 scored 1 point when the spinner landed on K. Player 2 scored 1 point when the spinner landed on J. Which spinner makes the game fair?

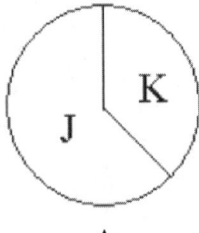   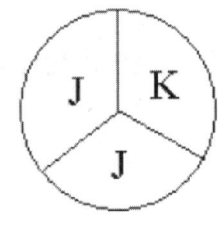   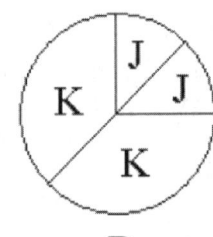

A.   B.   C.   D.

12. A bag contains four green marbles and eight red marbles. Bruce pulls out a marble. What color marble is Bruce most likely to pick? Explain why.

13. Jill is thinking of a number pattern. The first four numbers in her pattern are:

5, 10, 15, 20, _____, _____, _____ ...

Can the number 81 be in Jill's pattern? Explain your answer.

14. Complete the chart.

Favorite Color		
Color	Tally	Number
Red	ⅢⅡ	5
Blue	‖	2
Yellow	ⅢⅡ ‖‖	8
Green	ⅢⅡ ‖	6

a. How many people were surveyed? _21_

b. What was the most popular color? _yellow_

c. What was the least popular color? _blue_

Favorite Sport	
Sport	Number of Students
Football	🧍🧍🧍🧍
Soccer	🧍🧍🧍🧍🧍
Baseball	🧍🧍🧍
Basketball	🧍🧍🧍

Key: 🧍 = 4 students

15. a. How many students play soccer? _____ fiinty_____

b. How many students play baseball? _____

c. How many students play either football or baseball? _Sextin_

Questions 1 – 6 should be answered without a calculator.

1. $32 + 9 =$

    A. 39   B. 40   C. 41   D. 42

2. $45 \div 5 =$

    A. 5   B. 6   C. 8   D. 9

3. $60 - 18 =$

    A. 32   B. 38   C. 40   D. 42

4. Round 134 to the nearest ten.

    A. 100   B. 130   C. 135   D. 140

5. Which number is even?

    A. 25   B. 121   C. 503   D. 1600

6. One-tenth written as a fraction is

    A $\frac{1}{2}$ B. $\frac{1}{4}$ C. $\frac{1}{10}$ D. $\frac{1}{100}$

7. Tom ate 4 slices of a pizza that had been cut into 8 equal slices. What fraction of the pizza was left?

A. $\dfrac{5}{8}$ B. 5   C. $\dfrac{4}{8}$ D. $\dfrac{8}{8}$

8. The video store rented out 379 movies on Saturday and 487 movies on Sunday. Estimate the number of movies rented that weekend.

A. fewer than 300   C. between 500 and 700

B. between 300 and 500   D. more than 700

9. Apples are 25¢ each. Oranges are 4 for $1.00. Using this information, what question can you <u>not</u> answer?

A. How much does one orange cost?

B. How much does it cost to make orange juice?

C. How much do 4 apples cost?

D. How much does an apple and an orange cost?

10. Pencils cost 14¢ each. Anthony bought 3 pencils. He gave the cashier $1.00. What operations would you use to find out how much Anthony will get back?

A. subtract, then add

B. divide, then subtract

C. multiply, then subtract

D. multiply, then add

11. Marshall is 10 years old. Dale is 3 years younger than Marshall. Ron is 2 years younger than Dale. How old is Ron?

A. 5   B. 6   C. 7   D. 8

12. If you buy 4 pencils for $1.00, how much will it cost to buy 12 pencils?

A. $2.00   B. $ 3.00   C. $ 8.00   D. $12.00

13. Kara, Melanie, and April are going to tour the Grand Canyon. They are taking a small plane and have to give their weight. Kara weighs 102 pounds, Melanie weighs 147 pounds, and April weighs 98 pounds.

Estimate the combined weight of the three girls.

If the plane can only carry a total of 300 pounds, can they all go on the tour? no

Explain your answer.

The girls cant go on the plane because the plane can carry 300 not 350

14. Using the digits 6, 5, and 4, what is the largest possible odd number?

   A. 645   B. 654   C. 456   D. 564

15. There are 26 students in Mrs. Vanderlock's class.
   If 19 students are boys, how many girls are in the class?

   A. 6   B. 7   C. 8   D. 17

16. Doug wants to buy some tools. The prices of the tools are:

   $ 3.99   Hammer $ 1.99   Pliers
   $ 2.99   Wrench $ 0.99   Ruler
   $ 0.99   Screwdriver

   List 2 different combinations of tools Doug can buy with $8.00.

   Hammer   _Ruler_   _Pliers_

   Hammer   _wrench_   _Ruler_

17. Last year, there were 41 days of snow, 21 days of rain, and 69 cloudy days. On all the other days the weather was sunny.
   Using 365 days for the whole year, how do you calculate the number of sunny days?

   A. 365 – 41 – 21 – 69
   B. 41 + 21 + 69
   C. 365 + 41 + 21 + 69
   D. 41 + 21 + 69 - 365

158

18. Mariah practices the violin for 45 minutes. She finished practicing at 6:15 P.M. What time did she start to practice?

   A.   4:45 P.M.   C.   5:30 P.M.

   B.   5:00 P.M.   D.   5:45 P.M.

19. Which of these is heavier than a pound?

   A.   a crayon   C.   a bed

   B.   a pencil   D.   a sock

20. Nick bought a notebook for $1.73. He gave the cashier $2.00. How much money does Nick get back?

   A. $ 0.25   C. $ 0.32

   B. $ 0.27   D. $ 0.33

21. Mrs. Winters bought a can of peas for $0.48. What combination of coins could she get as change if she gave the cashier a dollar bill?

   Explain your answer.

   2 quarters 2 pennies
   25   10 nickels   2 pennies
   x4   5 dimes   2 pennies
   100
   -48

22. What shape (solid figure) best describes a can of soup?

A. cube   C. sphere

B. cone   D. cylinder

23. Find the perimeter of the figure shown.

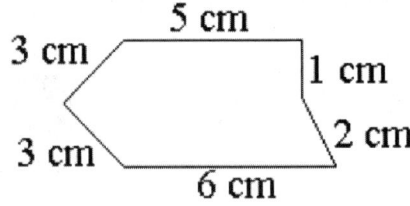

A. 11 cm   C. 18 cm

B. 26 cm   D. 20 cm

24. Which of the numbers has a 5 in the tens place and a 2 in the thousands place?

A. 3,250   B. 2,350   C. 5,203   D. 2,503

25. Draw a line of symmetry for each shape below.

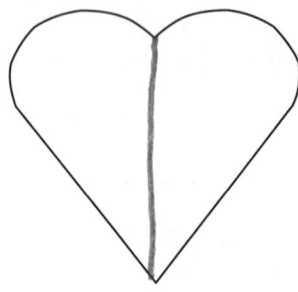 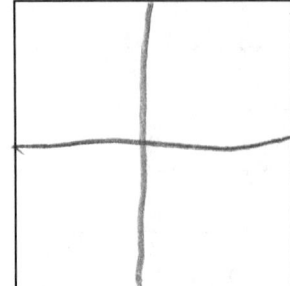 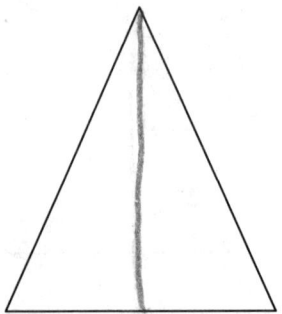

Soccer Goals Scored

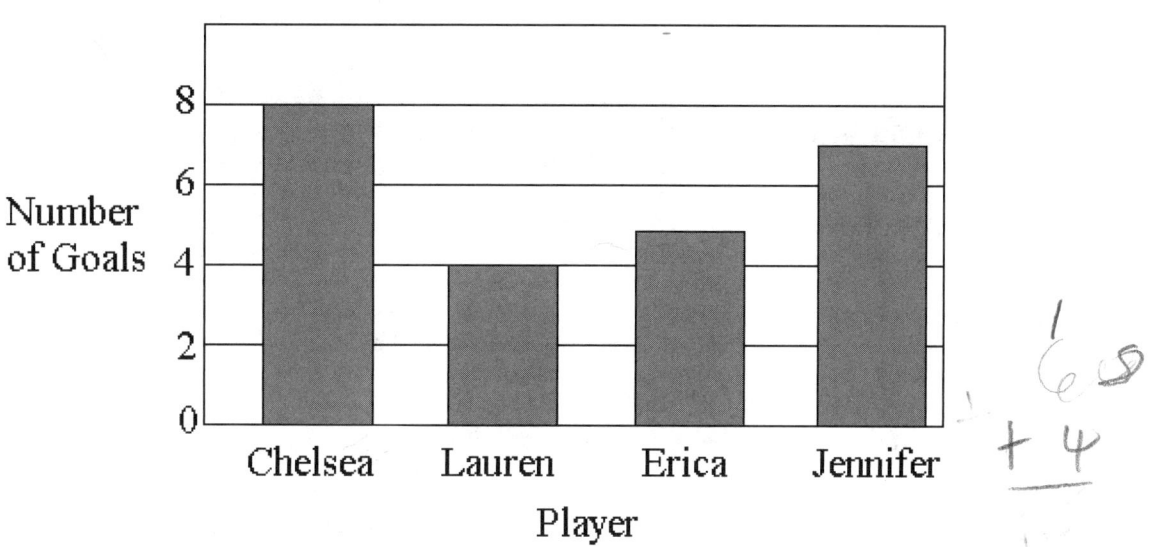

26.   How many goals did Erica have?   _5_

How many goals did Chelsea and Jennifer have altogether?   _15_

Who had the least number of goals?   _Lauren_

What is the total number of goals scored by all girls?   _24_

Favorite Holidays

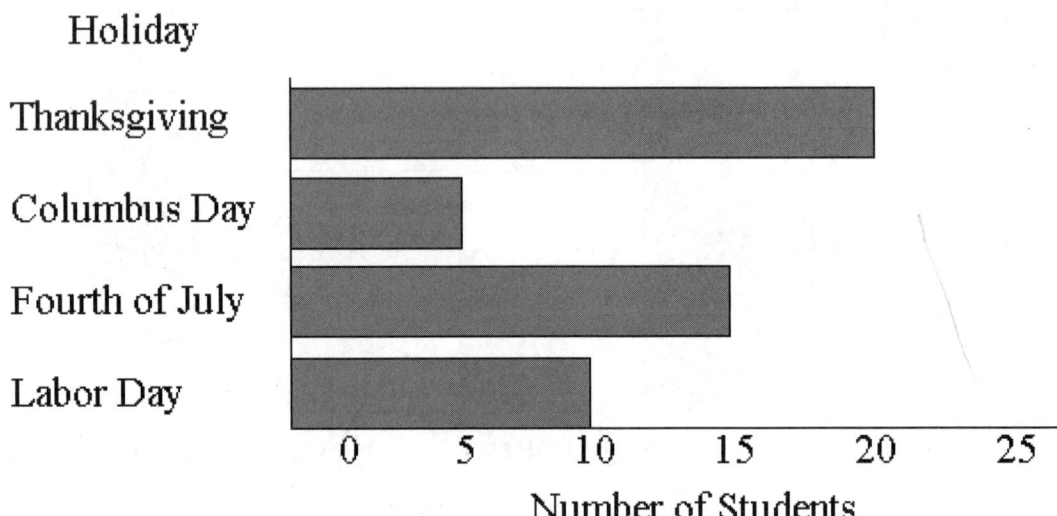

27. How many students preferred each of the following holidays:

Thanksgiving _____ 20

Columbus Day _____ 5

Fourth of July _____ 15

Labor Day _____ 10

Which was the most favorite holiday? _____
Explain your answer.

Thanksgiving becaue the students pick it.

_____

_____

_____

28. Francisco played three games of basketball. He scored 8 points during his first game, 22 points during his second game, and 38 points during his third game.
Estimate the total number of points.

A. Fewer than 50
B. Between 50 and 60
C. between 60 and 70
D. more than 80

29. One quart of milk makes four cups. How many quarts of milk do you need to give a cup of milk to each of 12 students?

A. 2   B. 3   C. 4   D. 12

30. Which numbers will complete the pattern?

12, 15, __, 21, __, 27, 30.

A. 17 and 23
B. 17 and 24
C. 18 and 24
D. 18 and 25

31. I have 7 coins and the total is $1.25. What are the coins?

A. 4 quarters, 2 dimes, 1 nickel.
B. 5 quarters
C. 4 quarters 2 dimes 5 pennies
D. 4 quarters and 5 nickels

End of Practice Test 1

Practice Test 2

Questions 1 – 6 must be answered without a calculator.

1. 34 + 16 =

    A. 40   B. 50   C. 56   D. 60

2. 6 x 5 =

    A. 11   B. 25   C. 30   D. 35

3. 19 – 10 =

    A. 9   B. 10   C. 11   D. 12

4. Estimate the sum of 68 and 121 = _____

    A. 160   B. 170   C. 180   D. 190

5. 222 ÷ 2 =

    A. 12   B. 22   C. 111   D. 101

6.   185 x 0 =

   (A.) 0   B. 184   C. 185   D. 186

7.   What is the fewest number of coins needed to make 66¢, using
      pennies, nickels, dimes, and quarters?

      A.   4   (B.) 5   C.   7   (D.   8

8.   Using three of the digits 0 through 9 only once, what is the largest
      3-digit even number you can make?

      _____   _____   _____

      What is the largest 3-digit odd number you can make?

      _____   _____   _____

9.   How many numbers between 70 and 80 are odd?

      A.   4   B.   5   C.   6   (D.) 7

10. Which of the following shows the numbers in order from least to greatest?

A.   67   7   76   C.   7   67   76

B.   67   76   7   D.   76   67   7

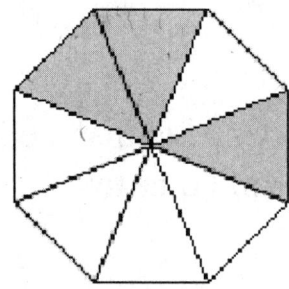

11. Choose the answer that represents the fraction for the shaded part of the above figure.

A. $\frac{5}{3}$ B. $\frac{3}{8}$ C. $\frac{8}{5}$ D. $\frac{5}{8}$

12. Susan looked at a thermometer in the winter and this is what she saw. What temperature does the thermometer show? (Hint: Skip count by 2's.)

A. -32°
B. -31°
C. 31°
D. 32°

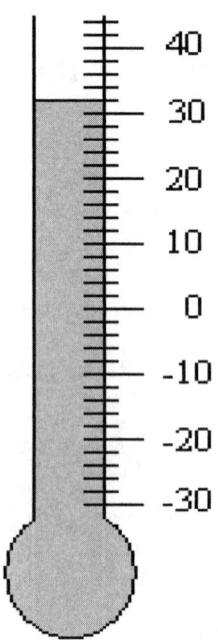

13. There were 96 adults at the zoo on Monday and 82 adults at the
     zoo on Tuesday. How many more adults were at the zoo on
     Monday than on Tuesday?

     A. 14   B. 18   C. 82   D. 96

14. Look at the spinner. Write the correct answer.

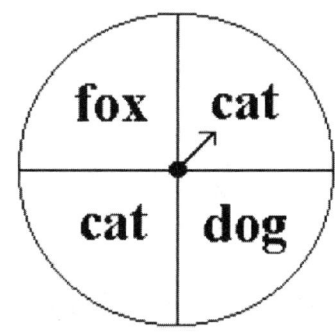

     What are the chances (probability) of the pointer   _____
     stopping on fox?

     What are the chances (probability) of the pointer   _____
     stopping on cat?

     On which animal is the pointer most likely to stop?   _____

     _____

15 . Which number sentence is true?

     A. 647 < 576   C. 546 > 664

     B. 674 < 647   D. 486 > 457

16. Ron bought a ball for 67 cents. He gave the cashier the exact amount of coins. How many coins did he give to the cashier?

 A   2 quarters, 2 dimes, 2 pennies
 B. 2 quarters, 2 dimes, 7 pennies
 C. 1 quarter, 3 dimes, 1 nickel, 2 pennies
 D. 2 quarters, 1 dime, 1 nickel, 2 pennies

17. $60 \times 40 =$

 A   24   B.   240   C.   2,400   D.   24,000

18. Write the next five numbers that continue this pattern.

   21, 22, 25, 26, 29, 30, _33_ _34_ _35_ _38_

 A. 33, 34, 35, 38, 39
 B. 31, 32, 33, 34, 35
 C. 32, 33, 36, 37, 40
 D. 33, 36, 39, 42, 45

19. Each face of a number cube is labeled with one of the numbers 1 through 6. What is the probability of rolling a 1?

 A. $\dfrac{1}{6}$ B. $\dfrac{1}{3}$ C. $\dfrac{1}{2}$ D. 1

20. Which pair of numbers should come next in the column?

x	6	7	8	?
y	36	42	48	?

A. (8, 54)   B. (8, 56)   C. (9, 54)   D. (9, 56)

21. Brandon looked at a thermometer at 2:00 P.M. on a winter day and this is what he saw.

By 4:00 P.M. the sun came out, and it warmed up by 10 degrees. What was the temperature at 4:00 P.M.?

A. - 12° C
B. - 4° C
C. 4° C
D. 6° C

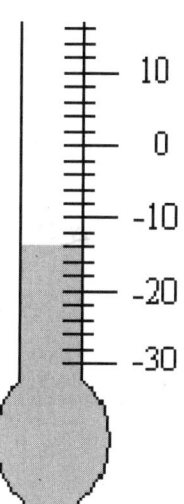

22. Which word form matches this numeral: 4,205

   A. four hundred twenty-five
   B. four thousand twenty-five
   C. four thousand two hundred and five
   D. four thousand two hundred zero five

23. The newspaper reported the following temperatures for Central New Jersey during a certain week in August:

Day	Temperature (degrees Fahrenheit)
Monday	85°
Tuesday	78°
Wednesday	83°
Thursday	88°
Friday	81°
Saturday	66°
Sunday	65°

What was the median temperature of the week?

A.   75°   C.   81°

B.   78°   D.   83°

24. Emma drinks 2 glasses of orange juice every weekday, and 1 glass on each Saturday and Sunday.
How many glasses of orange juice does Emma drink in one week?

A. 7   B. 10   C. 12   D. 14

25. The graph below shows the attendance for the first three days of the fireman's carnival.

Attendance at Fireman's Carnival

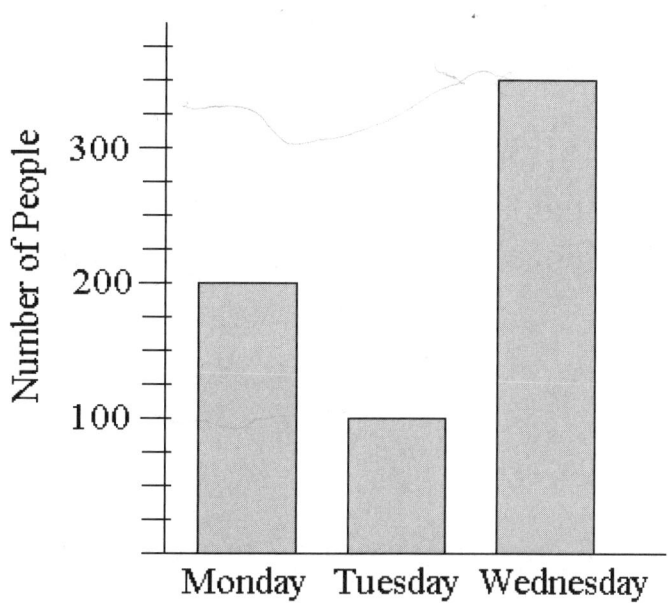

On which day did the greatest number of people attend?

wednesday

On which day did the least number of people attend? _Tuesday_

How many people attended on Monday? _200_

How many people attended on Tuesday? _100_

How many people attended on Wednesday? _350_

$$\begin{array}{r} 300 \\ + 50 \\ \hline 350 \end{array}$$

26. I have 7 coins that total $1.00. What are the coins?

    A. 2 quarters, 5 dimes
    B. 3 quarters, 2 dimes, 2 nickels
    C. 1 quarter, 6 dimes
    D. 2 quarters, 4 dimes, 1 nickel

27. Adam has a rectangular garden that is 3 feet long and 2 feet wide. How long is the fence that goes all around his garden?

    A. 3 feet   B. 5 feet   C. 6 feet   D. 10 feet

28. Which is less: 16 cups of water or 2 pints of water?

    Explain your answer.

    _____

    _____

    _____

    _____

    _____

    _____

    _____

    _____

29. Carlos has basketball practice at 6:00 PM. It is one hour and twenty minutes long. At what time will Carlos be finished?

    A. 7:20 P.M.   C. 8:00 P.M.
    B. 7:40 P.M.   D. 8:20 P.M.

30. Catherine's grandmother is 75 years old. Catherine is 7 years old. How much older is Catherine's grandmother?

    A. 65 years   C. 68 years
    B. 67 years   D. 70 years

31. Clayton ran 1 mile on Friday, 1.5 miles on Saturday, and 2 miles on Sunday. How many miles did he run altogether?

    A. 4 miles   C. 5 miles
    B. 4.5 miles   D. 35 miles

End of Practice Test 2

1. Dana has 15 baseball cards. Her brother Warren gave her 15 more baseball cards. How many does she have now?

    A. 16   B. 25   C. 30   D. 35

2. In the football game, Alan scored 12 points. Then he scored 21 more points. What is the total number of points he scored?

    A. 23   B. 33   C. 41   D. 43

3. Beth picked three baskets of tomatoes. Each basket had 14 tomatoes. How many tomatoes were there in all?

    A. 14   B. 28   C. 39   D. 42

4. Jay buys a sandwich for $3.55. He pays for it with a five dollar bill. How much change does he receive?

    A. $ 1.45   B. $ 2.45   C. $ 2.55   D. $3.45

5. Miles and Darren have 16 coins total. Miles has 2 more coins than Darren. How many coins does Darren have?

    A. 6   B. 7   C. 8   D. 9

6. Stan has 32 stamps. He bought 29 more. Stan wrote 61 letters and put stamps on them. How many stamps does he have left?

7. Patrick saves $5.00 each week. At the end of week 1, he saved $5.00. At the end of week 2, he saved $10.00. At the end of week 3, he saved $15.00. How much will Patrick save at the end of week 5 if he continues this pattern?

8. Joshua has cars on a shelf. Three cars are on the first shelf, four cars are on the second shelf, six cars on the third shelf and nine cars on the fourth shelf. How many cars will be on the fifth shelf if the pattern continues?

9. Stephen took 32 pages of notes. He had 150 pages in his notebook. How many pages were left in Stephen's notebook?

10. The third grade is taking a field trip. The bus holds 54 students. Ms. Baron has 24 students, and Ms. Jacobs has 23 students. How many students will be going on the trip?

11. I am the greatest two digit number. If you round me to the nearest ten, I become 100. What number am I?

12. I am a three digit number. I am an odd number. If you round me to the nearest ten, I become 500. The sum of my digits equals the number 8. What number am I?

13. George collects coins. He has 256 coins in all. How many coins has George collected, rounded to the nearest ten? Show or explain your answer.

14. Will's race boat must weigh 750 pounds with the driver in it. Will's boat weighs 622 pounds, and Will weighs 103 pounds. How much weight does Will need to add to his boat for the total weight to be 750 pounds?

15. Ricco scored 85 points in his first race and 80 points in his second race on Saturday. On Sunday he scored 90 points in his first race and 95 points in his second race. How many points did he earn all together?

16. Holly will baby-sit for four hours. She is paid $5.00 an hour. How much money will Holly make baby-sitting?

17. Ms. McCabe's class has 24 students. Half of her class are boys. How many are girls?

18. Valerie lives in an apartment building. She lives 3 floors above Dominique. Dominique lives on the 5th floor. What floor does Valerie live on? Show and explain your answer.

19. Vanessa wrote the pattern 14, 24, 34, 44. What should be the next two numbers in the pattern?

20. Jaylene is walking down the street. The addresses on the houses are 651, 653, 655, _____, 659. What address number is missing?

21. The lockers on the third floor are numbered 12, 14, 16, and 18. If the pattern continues, what will the next locker number be?

22. Ms. Bennington drove 2,496 miles to visit her parents. Rounded to the nearest thousand, how many miles did Ms. Bennington drive?

23. Alicia bought a hat for $6.73. She paid for it with a ten dollar bill. How much change did she receive?

24. Catherine is buying school supplies. She pays $1.99 for pencils and $1.79 for paper. She will pay for it with a five dollar bill. How much change will she get back?

25. Jordan went shopping for gifts for his family. His total bill was $97.58. He did not have enough money to pay his bill. He borrowed $60.00 from his mother. How much money did Jordan have of his own?

26. Laura has 4 dollars, 2 quarters, 3 dimes, and 2 pennies. How much money does Laura have?

27. Cheryl has $44.35 in her piggy bank. How much will she have if she puts 1 dollar, 2 quarters, and 1 dime in her piggy bank?

28. Jose, Ellen, and Millie bought cookies from the school bake sale. Jose spent $1.65, Ellen spent $0.65, and Millie spent $2.55. Put the amounts in order from greatest to least. Who spent the most?

29. Judy bought a bag of popcorn at the movies for $2.25. She paid with a five dollar bill. How much change will she get back?

30. Christopher bought a ticket to the movies for $5.00. He bought a soda for $1.25. What is the total amount he spent at the movies?

31. There are 235 students in the Roxbury school choir and 168 in the school band. How many students are in the choir and band all together?

32. Jodi bought a birthday card for $2.75 and wrapping paper for $2.98. What was the total cost of the card and wrapping paper?

33. Mr. Ramos drove 16 miles to school. Then, he drove 28 miles to the lake. Finally, he drove 38 miles to get back home. How many miles in all did Mr. Ramos drive that day?

34. Debbie had a basket of flowers. She had 29 flowers in the basket. Debbie picked 16 more flowers and put them in the basket. How many flowers are now in the basket?

35. Nancy had 61 stamps. She put the stamps on 55 envelopes to be mailed. How many stamps does she have left?

36. There are 33 students on Bus #77 going to school in the morning. Only 15 students take the bus home in the afternoon. Write a subtraction number sentence to find out how many more students take the bus to school in the morning.

37. Zach drives with his grandparents to Orlando, Florida. The trip is 916 miles long. They drove 301 miles before stopping for lunch. Estimate how many more miles they have to reach Orlando?

38. A boat company built 284 boats. They painted 84 white. How many more boats still need to be painted?

39. The New Jersey school year has 181 days. In Japan, the school year has 251 days. How many more days do Japanese children go to school each year?

40. Angelina bought 3 boxes of candy. Each box cost $3.00. How much did the candy cost altogether?

41. A dog has 4 paws. How many paws will 4 dogs have?

42. Harry has a remote control car that will run for 10 miles per hour. If Harry runs his car for 4 hours, how many miles will his car run?

43. You have 5 boxes of crayons. Each box has 10 crayons. How many crayons do you have?

44. Adam is making omelets. Each omelet needs 3 eggs and cheese. How many eggs are needed for 4 omelets?

45. Julian read 4 pages in his book each day for 7 days. How many pages did he read in all?

46. Ricky's birthday is 3 weeks away. How many days is it until Ricky's birthday? (Hint: How many days are in a week?)

47. An octopus has 8 legs. How many legs do 6 octopuses have?

48. Theresa has 20 cookies. She wants to share the cookies with 3 of her friends. How many cookies will each friend receive?

49. Joan has 10 fish evenly divided into 2 bowls. How many fish were in each bowl?

50. The school library has 6 tables. There are 4 chairs at each table. How many chairs are in the library?

51. Mrs. Kenner's class has 30 students. She wants to make 6 equal rows of desks. How many desks will be in each row?

52. Ashley is making beaded necklaces. She puts 10 beads on each necklace. How many necklaces can she make with 90 beads?

53. There are 15 bottles of fruit juice in the cooler. If Salvatore drinks 3 of them a day, how many days will it take him to drink them all?

54. The school has 8 basketballs. Two students share a basketball. How many students are playing with the basketballs?

55. Tina made a pie. She served $\frac{4}{8}$ of the pie to her friends.

Draw a picture to show the pie that Tina made.

Shade the fraction of the pie that Tina shared with her friends.

56. Ryan and his mother baked a chocolate cake. Ryan ate one-third of the cake. His mother ate one-third of the cake.

What fraction of the cake did they eat?

What fraction of the cake was left?

57. In the Lincoln School there are 5 staircases, each of which has 20 stairs in it. How many stairs are in the Lincoln School?

58. Seth found 5 nickels. How much money did he find?

59. An egg carton has 2 rows of eggs with 6 eggs in each row. How many eggs are in the carton?

60. Jack's math book has 756 pages. His science book has 562 pages. How many pages are in both books?

61. The kitchen table is 4 feet wide and 6 feet long. What is the perimeter of the table?

62. Frank spent $5.86 on his lunch. Maureen spent $4.19 on her lunch.

Who spent more?

How much more?

63. Dylan is buying lunch for himself and two of his friends. Dylan's lunch cost $6.95, his friend Darrell's lunch cost $6.29, and his friend Aaron's lunch cost $5.79. How much did Dylan spend on lunch?

64. Luke went to the store with $4.95 in his pocket. He bought a loaf of bread for $1.98. How much money did he have left?

65. In the fall, Farmer A and Farmer B sell pumpkins. One the first day, Farmer A and Farmer B each sold one pumpkin. The next day Farmer A sold 2 pumpkins and Farmer B sold 4. On the third day, Farmer A sold 4 pumpkins and Farmer B sold 7. On the fourth day Farmer A sold 7 and Farmer B sold 10. If Farmer A and Farmer B keep selling pumpkins the same way, how many pumpkins will Farmer A sell on the day Farmer B sells 19?

66. Angela and Marion are making a fruit salad. They have 48 grapes. if they want to make 8 equal servings, how many grapes will be put in each fruit salad?

67. Megan waters her plants every week. It takes 10 minutes to water all the plants. In 8 weeks, how many minutes does she spend watering her plants?

68. Dennis wants to divide 49 balloons equally among 7 people. What multiplication fact can help him solve his problem? Show and explain your answer.

69. Mickey is making a pattern using his colored trains. The first five train cars are blue, red, green, and then blue, red. If he continues the pattern, what colors would the next two train cars be?

70. Peter drew a figure that has two pairs of sides that are equal. Draw the figure that Peter drew.

71. Keith cut a pizza pie into 8 equal pieces. He ate 2 pieces. What fraction of the pizza pie is left?

72. Albert is reading a book that has 100 pages. He has read 28 pages. Write the fraction and decimal that shows the part of the book that Albert has read.

73. There are 10 textbooks on the shelf in Mrs. Chow's classroom. Seven of these books are math textbooks. Write the fraction and decimal for the number of math textbooks.

74. Cheyenne buys 6 bottles of juice. Each bottle holds 32 ounces. How many ounces of juice did Cheyenne buy?

75. Bobby has a paper route.
- He has more than 50 and less than 70 customers.
- The two digits in the number of customers he has are 1 apart.
- The sum of the two digits is less than 10.

How many customers does Bobby have on his paper route?

76. Which of these numbers has a 5 in the tens place, 2 in the thousands place, and 3 in the hundreds place?

   A.   3,250   C.   2,350

   B.   2,503   D.   5,203

77. Donald and his family are going on vacation. They look at the chart to see how far they have to drive to get there.

- The distance is less than 300 miles.
- The distance doesn't have the digits 7 or 4 in it.
- The distance is an even number.

	Trenton	Buffalo	New York	Philadelphia	Washington	Richmond
Trenton	---	454	212	302	437	593
Buffalo	454	---	434	381	217	384
New York	212	434	---	93	384	227
Philadelphia	302	381	93	---	302	139
Washington	437	217	384	302	---	240
Richmond	593	384	227	139	240	---

Between what 2 cities are they traveling?

78. Francisco played 3 games of basketball. He scored 8 points his first game, 28 points his second game, and 56 points his third game. About how many total points did he score?

   A.   between 60 and 70   C.   between 80 and 90

   B.   between 70 and 80   D.   between 90 and 100

79. The clock shows the time Kim began soccer practice. Soccer practice runs for 1 hour and 45 minutes. What time did practice end?

   A.   6:00

   B.   4:45

   C.   5:00

   D.   4:15

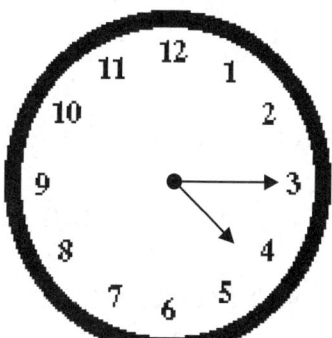

80. Shawn and Jeremiah went fishing. Shawn caught 6 fish, and Jeremiah caught 13 fish. Which number sentence can you use to find how many more fish Jeremiah caught?

   A.   13 − 6 = _____   C.   13 × 6 = _____

   B.   13 + 6 = _____   D.   13 ÷ 6 = _____

81. How many numbers in the box are even numbers?

782	229	448	962	627

   A.   three   C.   four

   B.   one   D.   two

82. There are 24 students in Ms. McCartney's class. If only one-third of the students completed their math project, how many students completed this project?

A. 8

C. 16

B. 12

D. 20

83. Ray bought a ball for 67 cents. He gave the cashier the exact amount of coins. What combination of coins could <u>not</u> be given to the cashier?

A. 2 quarters, 1 dime, 1 nickel, and 2 pennies

B. 5 dimes, 2 nickels, and 2 pennies

C. 4 dimes, 5 nickels, and 2 pennies

D. 1 quarter, 4 dimes, and 2 pennies

84. If $43 - \Box = 30$, what value goes in the box?

A. 13

C. 23

B. 17

D. 31

85. Which numbers will complete the pattern?

12, 14, 16, ___, 20, ___, 24

A. 18 and 22

C. 26 and 20

B. 17 and 21

D. 2 and 4

86. Jasmine is 5 years younger than Duke. Duke is 2 years older than Alex. Alex is 11 years old. How old is Jasmine?

A. 8

B. 12

C. 14

D. 18

87. What shape is located at the ordered pair (4, 5)?

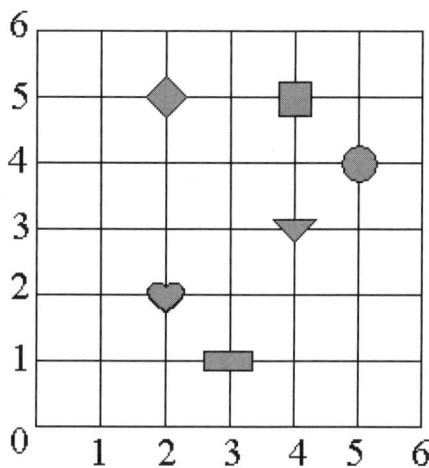

A. diamond

B. heart

C. square

D. circle

88. Which number sentence is <u>not</u> true?

A. 647 > 576

B. 674 > 647

C. 546 < 664

D. 457 > 486

89. Which letter has more than one line of symmetry?

A. A

B. H

C. V

D. M

90. How many numbers in the box become 400 when rounded to the nearest 100?

| 382 | 458 | 324 | 438 | 362 |

    A.   one   B.   three

    B.   two   D.   four

91. Carl had a vegetable garden that was a rectangular shape. It was 10 feet long and 5 feet wide. Carl wants to put a fence around his garden. How many feet of fencing will he need?

    A.   15 feet   C.   25 feet

    B.   20 feet   D.   30 feet

92. Lenny has 82 trading cards in his collection. His grandmother gave him more cards for getting a good report card. Now he has 106 cards. Which number sentence describes what took place?

    A. $82 + \_\_\_ = 106$ C. $82 - \_\_\_ = 106$

    B. $106 + 82 = \_\_\_$ D. $106 + \_\_\_ = 82$

93. What is another way to write six hundred eighty six?

    A.   6 hundreds 86 tens

    B.   60086

    C.   686

    D. $6 + 8 + 6$

94. How many sides does a cube have?

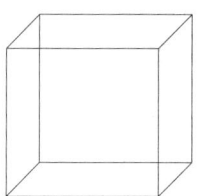

A. 1   C. 5

B. 3   D. 6

95. Which of these numbers is an even number greater than 535 and less than 591?

A. 596   C. 543

B. 562   D. 534

96. Elise needs a notebook that costs $5.99, a pen for $2.98, and a dictionary for $3.95. She gave the clerk $15.00. What operation would you use to find out how much change Elise will receive?

A. subtract, then add

B. add, then divide

C. add, then subtract

D. add, then multiply

97. What number goes in the box to make the number sentence true?

$$26 + 52 + 31 = 31 + 26 + \square$$

A. 26   C. 52

B. 31   D. 109

98. Hank has 3 blue gumballs, 2 red gumballs, and 4 yellow gumballs in a bag. What are his chances of selecting a red gumball?

    A. 1 in 2   C. 1 in 9

    B. 2 in 3   D. 2 in 9

99. Angelo works at the bakery. He sells different types of bread. Steven ordered two types of bread and paid Angelo $2.40. If Steven ordered onion bread, what other type of bread did he buy?

Bread Prices	
Onion	$1.15
Pizza	$1.25
Cheese	$1.50
Rye	$1.40
Sourdough	$1.45

    A. Pizza   C. Sourdough

    B. Cheese   D. Rye

100. What best describes the weight of a paper clip?

    A. gram   C. centimeter

    B. liter   D. kilogram

101. Chad is 6 inches shorter than Kurt. Kurt is 5 inches taller than Victor. Victor is 69 inches tall. How tall is Chad?

102. Eric's lucky number is greater than 18 and less than 27. It is an even number. You will say his number when you count by 3's. What is Eric's lucky number?

103. Joanne read 3 chapters in her book on Monday. On Tuesday she read 4 chapters and on Wednesday 5 chapters. If Joanne continues to increase her daily reading and she finishes the book on Friday, what is the greatest number of chapters the book can have?

104. Ali, Devon, Madeline, and Bethany are making up teams in gym. They want to have 2 people on each team. Show all the different ways that 2 people can be on a team using these 4 people?

105. Jonathan wanted to buy a box of crayons. He paid with 3 dimes, 2 nickels, and 4 pennies. How much did the crayons cost?

106. Juanita ordered a hamburger with lettuce, tomato, and onion. List all the different ways she can put these three items on top of her hamburger. Show or explain your answer.

107. Write clues so that the answers make a "number puzzle."

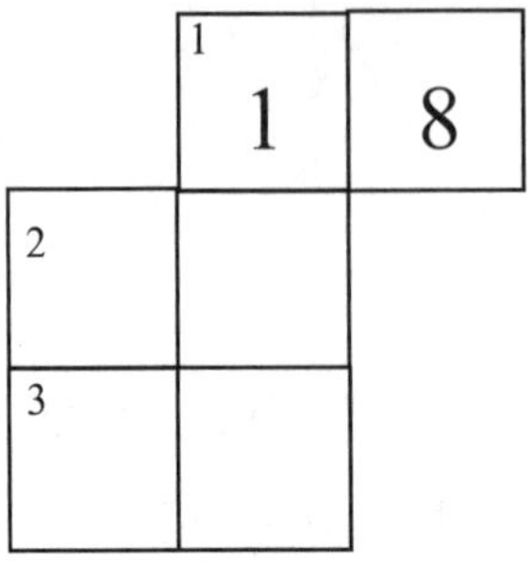

Across

1. 28 – 10 =

2.

3.

Down

1.

2.

Magic Shop	
Magic Cards	\$5.00
Magic Rings	\$10.00
Magic Hat	\$8.00
Magic Cloak	\$8.00
Magic Wand	\$2.00
Magic Rope	\$4.00
Magic Scarf	\$15.00

108. Adrian bought 3 items and spent exactly \$20.00 at the Magic Shop. What 3 items could Adrian have bought? Show or explain your answer.

109. The third grade class collected newspapers. The first week they collected 15 pounds of paper, the second week 21 pounds, and the third week 27 pounds. Which number sentence would you use to estimate the total number of pounds collected?

A. $20 + 20 + 30 = $ _____

C. $10 + 20 + 20 = $ _____

B. $20 + 10 + 20 = $ _____

D. $30 + 20 + 20 = $ _____

110. Mr. Bloom went to the town recycling center. He had 92 glass bottles, 56 aluminum cans, and 15 plastic milk containers. What number sentence could you use to estimate the total number of items he took to be recycled? Show or explain how you solved this problem.

111. Corey bought 3 boxes of cookies. There were 12 cookies in each box. What number sentence would you use to find the total number of cookies Corey bought?

A. $3 + 12 = $ _____

C. $12 - 3 = $ _____

B. $3 \times 12 = $ _____

D. $12 \div 3 = $ _____

112. Carrie had $20.00. She bought a shirt for $15.99 and received $4.01 as change. Which number sentence would you use to show this problem?

A. $\$20.00 - \$4.01 = \$15.99$

C. $\$4.01 - \$15.99 = \$20.00$

B. $\$15.99 - \$4.01 = \$20.00$

D. $\$20.00 - \$15.99 = \$4.01$

113. Complete this problem:

$$
\begin{array}{r}
4\ 7 \\
+\ \boxed{} \\
\hline
7\ 0
\end{array}
$$

114. Which number sentence is not true?

 A. $546 > 527$ C. $286 < 268$

 B. $826 > 628$ D. $736 < 763$

115. Which number would complete the number sentence to make it true?

$$3 \times \underline{\hspace{2cm}} > 2 \times 5$$

116. What number goes in the box to make the sentence true?

$$25 + 13 + 38 = \boxed{} + 25 + 13$$

 A. 23   C. 13

 B. 38   D. 15

117. The temperature in the morning was 69° F. The temperature in the afternoon was 91° F. Which number sentence shows the best way to estimate how many degrees the temperature changed.

 A. $70 + 100 = 170$ C. $100 - 70 = 30$

 B. $70 + 90 = 160$ D. $90 - 70 = 20$

118. Which bracelet does not show a line of symmetry?

A.

B.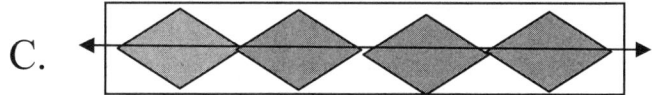

C.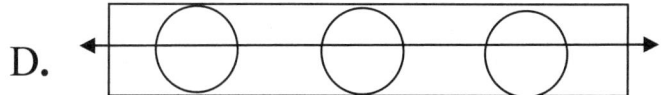

D.

119. The smallest number in a set of data is 41 and the largest number is 76. Which number sentence shows the best way to estimate the difference between both numbers?

A. $70 - 50 = 20$ C. $80 - 40 = 40$

B. $70 - 40 = 30$ D. $80 + 40 = 120$

120. Michelle drove 40 miles to Sandy Hook Beach and then drove to Paterson. She drove 70 miles all together. What number sentence can be used to find the distance Michelle drove from Sandy Hook Beach to Paterson.

A. $70 - 40 = 30$ C. $70 \times 2 = 140$

B. $35 + 35 = 70$ D. $35 + 70 = 105$

121. Ann arranged her Dominoes in the pattern shown below.

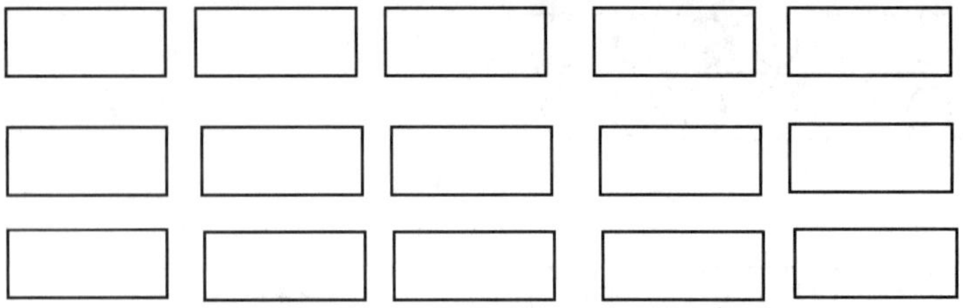

What answer choice best shows how many dominoes she arranged?

A. 3 + 5   C. 3 x 5

B. 5 + 5   D. 5 x 5

122. What answer best describes the pattern below?

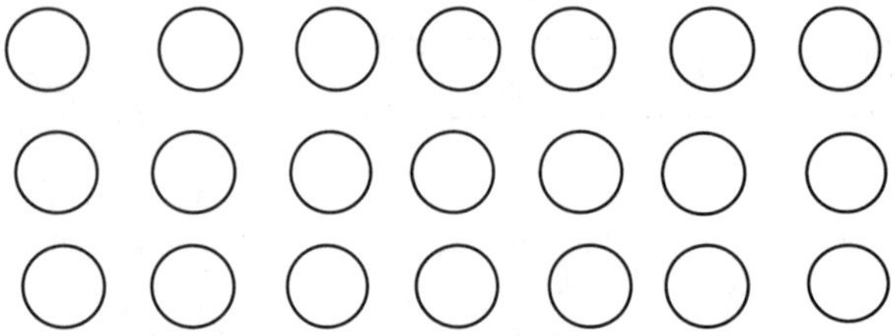

A.   3 x 7   C.   3 + 7

B.   7 x 7   D.   7 + 3